Medical Application of Functional Biomaterials

Medical Application of Functional Biomaterials

Guest Editors

Cristian Scheau
Andreea Cristiana Didilescu
Constantin Caruntu

MDPI

Basel • Beijing • Wuhan • Barcelona • Belgrade • Novi Sad • Cluj • Manchester

Guest Editors
Cristian Scheau
Department of Physiology
The "Carol Davila" University
of Medicine and Pharmacy
Bucharest
Romania

Andreea Cristiana Didilescu
Department of Embryology
and Microbiology
Faculty of Dentistry
The "Carol Davila" University
of Medicine and Pharmacy
Bucharest
Romania

Constantin Caruntu
Department of Physiology
The "Carol Davila" University
of Medicine and Pharmacy
Bucharest
Romania

Editorial Office
MDPI AG
Grosspeteranlage 5
4052 Basel, Switzerland

This is a reprint of the Special Issue, published open access by the journal *Journal of Functional Biomaterials* (ISSN 2079-4983), freely accessible at: https://www.mdpi.com/journal/jfb/special_issues/madical_biomaterials.

For citation purposes, cite each article independently as indicated on the article page online and as indicated below:

Lastname, A.A.; Lastname, B.B. Article Title. *Journal Name* **Year**, *Volume Number*, Page Range.

ISBN 978-3-7258-4205-6 (Hbk)
ISBN 978-3-7258-4206-3 (PDF)
https://doi.org/10.3390/books978-3-7258-4206-3

Cover image courtesy of Cristian Scheau

Contents

Journal of *Functional Biomaterials*

MDPI

Editorial

Medical Application of Functional Biomaterials—The Future Is Now

Cristian Scheau [1,*], Andreea Cristiana Didilescu [2,*] and Constantin Caruntu [1,3,*]

[1] Department of Physiology, The "Carol Davila" University of Medicine and Pharmacy, 050474 Bucharest, Romania
[2] Department of Embryology, Faculty of Dentistry, The "Carol Davila" University of Medicine and Pharmacy, 050474 Bucharest, Romania
[3] Department of Dermatology, "Prof. N.C. Paulescu" National Institute of Diabetes, Nutrition and Metabolic Diseases, 011233 Bucharest, Romania
* Correspondence: cristian.scheau@umfcd.ro (C.S.); andreea.didilescu@umfcd.ro (A.C.D.); costin.caruntu@gmail.com (C.C.)

We live in unprecedented times. Technology and manufacturing capabilities worldwide are rapidly expanding as new devices, substances, and materials are developed to fulfill the needs of different areas of interest, including medicine. Functional biomaterials have flooded the medical field in multiple domains with numerous applications, from replacing tissues and organs to consolidating regional physiology by promoting bone mineralization, balancing regional flora, and facilitating regeneration [1–3].

Polydimethylsiloxane (PDMS) is a widely used polymer with excellent biocompatibility, stability, and mechanical properties. Miranda et al. have presented an excellent review of recent applications of PDMS, a walkthrough of the properties and manufacturing process of this polymer, as well as the various medical applications of this substance found in recent literature, such as its use as a blood analogue, in implant coatings, or in replicating the cardiovascular system [4]. The authors also considered the microfabrication process of PDMS, which was thoroughly described, and further provided the best methods to characterize this compound and test its properties according to manufacturing standards. Auliya et al. have tested PDMS for its potential use as a vitreous substitute in the human eye [5]. In this study, compounds with viscosities of 1.15 Pa.s, 1.17 Pa.s, and 1.81 Pa.s were created, and the authors evaluated the physical properties of the mentioned samples. Furthermore, in vitro tests were performed to measure the toxicity of this material using the Hen's Egg Test Chorioallantoic Membrane method for the first time. Ultimately, it was shown that PDMS is safe and may replace the vitreous humor in human eyes—a very encouraging finding.

Titanium–nickel alloys have long been used in medical devices and implants, with papers going back to the 1980s citing the use of TiNi alloys, especially in the dental field [6]. However, recent advances in the manufacturing process have led to the development of porous TiNi alloys through self-propagating high-temperature synthesis, a material able to sustain the attachment and growth of human cells with great biocompatibility while maintaining excellent mechanical properties [7]. Topolnitskiy et al. have tested the performance of these types of implants in the repair of the chest wall after resection for malignant tumors [8]. The authors concluded that the material is safe and has good functional properties using one-stop surgery.

Stimulation of the body's regenerative properties is essential in healing after implant placement, and using appropriate substances is essential in favoring the acceptance of the newly introduced biomaterials. Bleyan et al. have shown that osseodensification may be used as a predictor for immediate implant placement using allografts or alloplasts as fillers [9]. They also proposed a new classification of the molar socket that takes into consideration the septum width before instrumentation and can be applied to estimate the impact on implant stability; according to the authors, S-IV sockets have less than 2 mm of initial septum width and are counterindications for septum expansion with osseodensification. The study of Srisomboon et al. assessed the restorative properties of silver diamine nitrate (SDN) and fluoride (SDF) on dentin remineralization while accounting for their cytotoxicity to dental pulp [10]. The authors proposed SDN as a more cost-effective alternative and developed a study to measure its effectiveness compared to SDF. SDF was shown to be more efficient than SDN in increasing apatite formation while showing similar cytotoxic effects on dental pulp cells. However, both compounds showed comparable precipitation of silver salts in the demineralized dentin, occluding the dentinal tubules; therefore, SDN was proposed as a potential material for caries control.

On a similar topic, Covaci et al. tested various calcium-based materials in deep carious lesions to assess their potential in maintaining dental pulp vitality [11]. The study revealed that self-setting calcium hydroxide was superior in the preservation of dental pulp vitality compared to resin-modified, calcium-releasing compounds. In addition, the pH of the compound was shown to be a factor that can affect the dental pulp and trigger inflammation or necrosis. These results advance the knowledge base on the biomaterials used in direct pulp capping, as the ideal material is still being sought.

Controlling the local environment around implants, prostheses, or other medical devices is key to preventing adverse reactions such as inflammation, infection, or various immune responses. In their study, Chanachai et al. prepared an adhesive containing monocalcium phosphate monohydrate (MCPM) and nisin to be used in fixed orthodontic treatment [12]. The material is strong enough to prevent bracket debonding and releases calcium phosphate, encouraging mineralization and buffering the local acidity while showing antibacterial properties. After a thoughtful interpretation of their results, the authors have proposed updated models of their developed compound that might have increased effectiveness in all these regards.

Moreover, the importance of biocompatibility and molecular interaction is underlined by Tizu et al. in their paper, where they elaborate on the interaction between two different polymeric nanoparticles and oral cells, more specifically, stem cells of keratinocyte as well as human exfoliated deciduous teeth [13]. The authors demonstrate the baseline for the polymer's uptake and note various significant parameters of the process. The findings are very relevant in terms of developing novel methods for tissue regeneration with reduced toxicity and high efficacy.

This Special Issue showed that the research field of functional biomaterials is of tremendous interest and importance, with rapid evolution and numerous applications. For medical usage of new biomaterials, it is required that safety and biocompatibility standards are met alongside the demonstration of the superior physical and chemical properties of the compounds. Adequate integration of biomaterials in complex biological environments that are thoroughly regulated by a wide array of physiological mechanisms will always represent a challenge for future developers of compounds, alloys, and substances. In the very first editorial of this Journal, the Founding Editor-in-Chief recognized the incredible potential of the area of biomaterials, the increasing scientific interest, as well as the multidisciplinarity involved in the development of this field [14]. It is our hope that our Special Issue will help advance the current knowledge of the medical applications of functional biomaterials and that it represents a proper forum for the intersection and cooperation of the scientific and technological communities, as Prof. Francesco Puoci intended for this Journal.

Author Contributions: Conceptualization, C.S., A.C.D. and C.C.; writing—original draft preparation, C.S., A.C.D. and C.C.; writing—review and editing, C.S., A.C.D. and C.C. All authors have read and agreed to the published version of the manuscript.

Funding: This research received no external funding.

Data Availability Statement: Not applicable.

Conflicts of Interest: The authors declare no conflict of interest.

References

1. Riedel, S.; Ward, D.; Kudláčková, R.; Mazur, K.; Bačáková, L.; Kerns, J.G.; Allinson, S.L.; Ashton, L.; Koniezcny, R.; Mayr, S.G.; et al. Electron Beam-Treated Enzymatically Mineralized Gelatin Hydrogels for Bone Tissue Engineering. *J. Funct. Biomater.* **2021**, *12*, 57. [CrossRef] [PubMed]
2. Iovene, A.; Zhao, Y.; Wang, S.; Amoako, K. Bioactive Polymeric Materials for the Advancement of Regenerative Medicine. *J. Funct. Biomater.* **2021**, *12*, 14. [CrossRef] [PubMed]
3. Hamdan, N.; Yamin, A.; Hamid, S.A.; Khodir, W.; Guarino, V. Functionalized Antimicrobial Nanofibers: Design Criteria and Recent Advances. *J. Funct. Biomater.* **2021**, *12*, 59. [CrossRef] [PubMed]
4. Miranda, I.; Souza, A.; Sousa, P.; Ribeiro, J.; Castanheira, E.M.S.; Lima, R.; Minas, G. Properties and Applications of PDMS for Biomedical Engineering: A Review. *J. Funct. Biomater.* **2022**, *13*, 2. [CrossRef] [PubMed]
5. Auliya, D.G.; Setiadji, S.; Fitrilawati, F.; Risdiana, R. Physical Characterization and In Vitro Toxicity Test of PDMS Synthesized from Low-Grade D4 Monomer as a Vitreous Substitute in the Human Eyes. *J. Funct. Biomater.* **2022**, *13*, 3. [CrossRef] [PubMed]
6. Kimura, H.; Sohmura, T. Surface coating on TiNi shape memory implant alloys. *J. Osaka Univ. Dent. Sch.* **1987**, *27*, 211–223. [PubMed]
7. Yasenchuk, Y.; Marchenko, E.; Baigonakova, G.; Gunther, S.; Kokorev, O.; Gunter, V.; Chekalkin, T.; Topolnitskiy, E.; Obrosov, A.; Kang, J.-H. Study on tensile, bending, fatigue, and in vivo behavior of porous SHS–TiNi alloy used as a bone substitute. *Biomed. Mater.* **2021**, *16*, 021001. [CrossRef] [PubMed]
8. Topolnitskiy, E.; Chekalkin, T.; Marchenko, E.; Yasenchuk, Y.; Kang, S.-B.; Kang, J.-H.; Obrosov, A. Evaluation of Clinical Performance of TiNi-Based Implants Used in Chest Wall Repair after Resection for Malignant Tumors. *J. Funct. Biomater.* **2021**, *12*, 60. [CrossRef] [PubMed]
9. Bleyan, S.; Gaspar, J.; Huwais, S.; Schwimer, C.; Mazor, Z.; Mendes, J.J.; Neiva, R. Molar Septum Expansion with Osseodensification for Immediate Implant Placement, Retrospective Multicenter Study with Up-to-5-Year Follow-Up, Introducing a New Molar Socket Classification. *J. Funct. Biomater.* **2021**, *12*, 66. [CrossRef] [PubMed]
10. Srisomboon, S.; Kettratad, M.; Stray, A.; Pakawanit, P.; Rojviriya, C.; Patntirapong, S.; Panpisut, P. Effects of Silver Diamine Nitrate and Silver Diamine Fluoride on Dentin Remineralization and Cytotoxicity to Dental Pulp Cells: An In Vitro Study. *J. Funct. Biomater.* **2022**, *13*, 16. [CrossRef] [PubMed]
11. Covaci, A.; Ciocan, L.T.; Gălbinașu, B.; Bucur, M.V.; Matei, M.; Didilescu, A.C. Dental Pulp Response to Different Types of Calcium-Based Materials Applied in Deep Carious Lesion Treatment— a Clinical Study. *J. Funct. Biomater.* **2022**, *13*, 51. [PubMed]
12. Chanachai, S.; Chaichana, W.; Insee, K.; Benjakul, S.; Aupaphong, V.; Panpisut, P. Physical/Mechanical and Antibacterial Properties of Orthodontic Adhesives Containing Calcium Phosphate and Nisin. *J. Funct. Biomater.* **2021**, *12*, 73. [CrossRef] [PubMed]
13. Tizu, M.; Mărunțelu, I.; Cristea, B.M.; Nistor, C.; Ishkitiev, N.; Mihaylova, Z.; Tsikandelova, R.; Miteva, M.; Caruntu, A.; Sabliov, C.; et al. PLGA Nanoparticles Uptake in Stem Cells from Human Exfoliated Deciduous Teeth and Oral Keratinocyte Stem Cells. *J. Funct. Biomater.* **2022**, *13*, 109. [CrossRef] [PubMed]
14. Puoci, F. Salve Journal of Functional Biomaterials, ad maiòra! *J. Funct. Biomater.* **2010**, *1*, 1–2.

Journal of
*Functional
Biomaterials*

Article

Evaluation of Clinical Performance of TiNi-Based Implants Used in Chest Wall Repair after Resection for Malignant Tumors

Evgeniy Topolnitskiy [1,2], Timofey Chekalkin [1,3,*], Ekaterina Marchenko [1], Yuri Yasenchuk [1], Seung-Baik Kang [4], Ji-Hoon Kang [3] and Aleksei Obrosov [1,5]

1 Laboratory of Medical Materials, Tomsk State University, 634045 Tomsk, Russia
2 Department of Surgery, Siberian State Medical University, 634050 Tomsk, Russia
3 R&D Center, TiNiKo Co., Ochang 28119, Korea
4 Boramae Medical Center, Seoul National University Hospital, Seoul 07061, Korea
5 Department of Physical Metallurgy and Materials Technology, Brandenburg University of Technology, 03-046 Cottbus, Germany
* Correspondence: tc77@rec.tsu.ru

Abstract: In this study, we assessed the outcomes after surgical treatment of thoracic post-excision defects in 15 patients, using TiNi knitted surgical meshes and customized artificial TiNi-based ribs. Methods: Eight patients were diagnosed with advanced non-small cell lung cancer (NSCLC) invading the chest wall, of which five patients were $T_3N_0M_0$, two were $T_3N_1M_0$, and one was $T_3N_2M_0$. Squamous cell carcinoma was identified in three of these patients and adenocarcinoma in five. In two cases, chest wall resection and repair were performed for metastases of kidney cancer after radical nephrectomy. Three-dimensional CT reconstruction and X-ray scans were used to plan the surgery and customize the reinforcing TiNi-based implants. All patients received TiNi-based devices and were prospectively followed for a few years. Results: So far, there have been no lethal outcomes, and all implanted devices were consistent in follow-up examinations. Immediate complications were noted in three cases (ejection of air through the pleural drains, paroxysm of atrial fibrillation, and pleuritis), which were conservatively managed. In the long term, no complications, aftereffects, or instability of the thoracic cage were observed. Conclusion: TiNi-based devices used for extensive thoracic lesion repair in this context are promising and reliable biomaterials that demonstrate good functional, clinical, and cosmetic outcomes.

Keywords: non-small cell lung cancer (NSCLC); thoracic lesion; chest wall reconstruction; TiNi artificial rib; TiNi mesh implant

1. Introduction

Primary malignant tumors of the chest wall are rare. More often, a secondary lesion is observed in the form of a tumor of the chest wall structures in lung/breast cancer, or malignant tumor of the mediastinum, or as a result of metastatic kidney, prostate, or thyroid cancer [1–4].

Despite the significant progress made in various areas of oncosurgery, the surgical stage in treating chest wall tumors occupies the leading position [3–7]. Radical surgical intervention for chest wall tumors is typically accompanied by the lesion of osteochondral structures and the appearance of complex post-resection defects, which result in functional and aesthetic impairment. Progressive surgical and intensive care technologies have enabled extended combination surgeries in which not only the tumor-affected chest wall fragment, but also the adjacent anatomical structures involved are excised as a single piece [3,7]. After extensive resection of the chest wall, it is vitally important that it is simultaneously repaired, including restoring the osteochondral framework and the integrity of the integumentary tissues as well as maintaining the anatomical and physiological volume of the mediastinum and the pleural cavities.

Despite the improved surgical techniques including modern implants, the reconstruction of extensive thoracic defects, regardless of etiology, is still challenging even for high-skilled surgeons. A variety of techniques for surgical treatment of post-excision thoracic defects indicates that there is no versatility of the proposed methods. All of them are have certain disadvantages and risks. This requires thorough planning, input from other specialists, customization, and a multidisciplinary approach. To date, the global clinical practice dictates an attitude towards reconstructive procedures, which suggested a tolerable postoperative complication rate. The mainstay of chest wall surgical treatment clearly shows that to eliminate an extensive post-excision defect of the integumentary soft tissues, it is recommended to adhere to the use of non-free skin/fatty tissue, skin/muscle, and muscle flaps, as well as the greater omentum on a pedicle [1,3,8–10]. However, to reconstruct the sternocostal framework, the use of synthetic materials and implants comes to the fore. Biomaterials most often applied for these purposes are meshes and sheets made of vicryl, mersilene, teflon, polycaproamide, polypropylene, polytetrafluoroethylene, titanium, stainless steel, or combinations thereof [11–17]. Noted disadvantages of these implants include secondary wound infections in up to 6% of cases [18], inadequate thoracic cage rigidity, respiratory dysfunction, and seroma, which often leads to revision surgery and the removal of an inconsistent device. There are published reports on the use of sternal and costal endografts made from fluorine- or carbon-containing materials, and of vascular grafts [12,13]. The concerns typically encountered when considering these surgical procedures are related to late complications, side effects, and inadequate follow-up and physical rehabilitation, stemming from the incomplete restoration of the framework of the thoracic cage.

Some studies report that osteosynthesis using Ti plates (Synthes MatrixRib, Stratos system) can be considered an attractive alternative in reconstruction of chest wall defects [19,20]. Such an approach is believed to eliminate paradoxical movement of the chest wall and facilitate normal chest wall mechanics [21]. Titanium and its nontreated surfaces are known to be bioinert to a certain degree [22,23], allowing tissue integration and compatibility with CT scanning. The use of Ti rib systems seems to supplement surgical options when using the biomaterials mentioned above. Moreover, reports have appeared in recent years on the reconstruction of complex post-excision chest wall lesions using customized titanium 3D-printed constructs of various designs [14–17]. However, even in the simple breathing cycle, the implanted device may suffer from complex loading, including tension, bending, and torsion. Of course, a good device used for rib-plasty should mimic the anisotropic compliance of the chest wall and should demonstrate a tolerant long-term stress-strain behavior without impairment of the mechanical characteristics at higher loads. Considering a Ti-based device, especially in the case of a long or sophisticated implant, the mechanical characteristics of which differ from that shown by the rib to be substituted, we can assume that there is a noncoincidence in the context of elastic moduli, which would result in eventual implant-induced complications like pain, rejection, failure, fibrosis, and inflammation. Moreover, crucial issues and limitations regarding 3D-printed devices, such as cost, reproducibility of the microstructure and properties, and scalability of the fabrication processes to mass production levels, still remain.

In this regard, there has been a search for an adequate technique in plasty of this type of lesions. Obviously, the suggestion of a superior surgical option using *ad hoc*, versatile, and affordable devices strives to reduce the specific obstacles faced by existing procedures, making the latter routine and increasing the success rate even for low-skilled surgeons.

Porous and solid TiNi-based implants and their successful deployment in surgical treatments have encouraged insights for immediate and delayed rib-plasty in cancer patients. Experimental and clinical studies with promising results have demonstrated successful integration of TiNi implants with the formation of regenerated tissues, which anatomically and physiologically restore the injured area [24–30]. Recently, porous SHS-

TiNi alloys have been reported [31] to have some features that significantly distinguish them from those manufactured by other methods of powder metallurgy. It happens that the porous compound formation during the SHS reaction is accompanied by the genesis of bio-active nonmetallics and nanocrystalline, corrosion-proof, and amorphous superficial layers concealing the pore walls [32], which are of great interest for clinical applications. Moreover, the rheological similarity between the viscoelastic artificial TiNi sandwich and the rib imparts additional working benefits to this biomaterial. The distinctive feature of porous TiNi is conditioned by the lowest elastic modulus similar to that demonstrated by the bone tissue, whereas a rheological resemblance in terms of stress-strain allows the artificial rib to be congruentially deformed without rupture and delamination, passing through a million cycles, as studied in [30]. The rough hydrophilic surface of porous SHS-TiNi was reported to sustain cell adhesion, growth, and proliferation via a system of interconnected macro-/micropores and grooves [33–35]. A number of clinical cases have been published, describing the successful elimination of chest wall skeletal defects through the use of TiNi implants in the surgical treatment of malignant chest wall tumors [36–38].

Here, we report our clinical experience based on 15 patients in the combined use of TiNi devices in anterior chest wall reconstructions following mid/large thoracic excisions and evaluate the outcomes of surgical treatment of chest wall malignant tumors. The novel aspect of this work is that the surgical method of post-excision defect repair was performed using a proprietary approach and customized TiNi-based implants. This can be disseminated and recommended as a routine surgical procedure with negligible complications and higher success rates.

2. Materials and Methods

Between May 2013 and December 2020, 15 patients (4 female and 11 male) with malignant tumors or tumor-like mass of the chest wall, invading the osteochondral structures, underwent surgery. The mean patient age was 60 years (range 26 to 73 years). Prior to surgery, the diagnostic standard of testing established for the specific disease was followed. The postoperative follow-up protocol for all patients included computed tomography (CT) of the chest with intravenous contrast, and magnetic resonance tomography for tumors located in the superior thoracic aperture and for tumors suspected of invading the spine. After multiplanar reconstruction of the patient's CT scans, an anatomical 3D model of the area was created that was used to plan the resection and reconstruction stages of the upcoming surgery and to customize the sizes/shapes of the TiNi-based implants.

The morphological variants of the primary and secondary chest wall tumors identified once the surgical specimens were examined and verified are summarized in Table 1. In the majority of cases (53.3%), the chest wall tumors were classified as advanced NSCLC with invasion of the chest wall; five patients were staged $T_3N_0M_0$, two were $T_3N_1M_0$, and one was $T_3N_2M_0$. Of the cases of NSCLC with chest wall invasion, the tumor was verified to be squamous cell carcinoma in three patients (37.5%) and adenocarcinoma in five (62.5%). Upon examination, five patients were found to have paracancerous inflammatory complications of lung cancer. In two cases, chest wall resection and simultaneous repair were performed for isolated metastases of renal cell cancer detected 18 and 24 months after radical nephrectomy. In one patient, the chest wall tumor was considered to be an isolated metastasis of adenocarcinoma of the lung into the anterolateral section of the left fourth rib 12 months after an extended right lower lobectomy. The patient had previously been treated with surgery and adjuvant chemotherapy complying with the standard EP regimen. In one case, a patient with breast cancer had undergone a radical right mastectomy and chemoradiotherapy, which was complicated by osteomyelitis of the sternum and ribs. After a series of curative surgical procedures, she developed an extensive chest wall defect combined with a ventral hernia and instability of the thoracic cage. At the final stage, once the inflammatory process had been addressed, the plastic correction procedure of the complications was performed.

Table 1. Patient distribution by morphological form of chest wall tumor.

Morphological Type	No. of Patients	
	Abs.	%
* NSCLC with invasion into the chest wall	8	53.3
Metastasis of renal cancer	2	13.3
Metastasis of NSCLC after radical lower lobectomy	1	6.7
Breast cancer	1	6.7
Plasmacytoma	1	6.7
Fibrous dysplasia	2	13.3
Total	**15**	**100**

* NSCLC-non small cell lung cancer.

In a complete physical examination of 14 (93.3%) of the patients, one or more concomitant diseases were identified (Table 2).

Table 2. Concomitant diseases in patients included in this cohort study.

Concomitant Diseases	No. of Patients	
	Abs.	%
Chronic obstructive pulmonary disease	9	64.3
Chronic nonspecific lung disease	1	7.1
Coronary artery disease	2	14.3
Abnormal cardiac rhythm	2	14.3
Type 2 diabetes mellitus	2	14.3
Obesity	2	14.3
Gastric and duodenal ulcer disease	1	7.1

To evaluate the postoperative status of each patient, the Charlson comorbidity index was used, with a mean score of four points. The physical status of patients by the American Society of Anesthesiologists classification was determined to be ASA II in 60% of the cases and ASA III in the remainder.

At the surgical stage of treatment, the sternal body and xiphoid process and the anterior sections of left ribs 5–9 and right ribs 5–7 were excised in one case; anterolateral sections of four ribs were excised in two cases, three ribs in five cases, and two ribs in five cases; and of the latter five cases, posterolateral sections of ribs were excised in two cases and the anterolateral section of one rib was excised in two cases. Besides resecting anatomical structures of the chest wall, one case had an atypical resection of the upper lobe of a lung, eight had extended upper lobectomies, and one had a left pneumonectomy. In one case, when fibrous dysplasia with osteolysis of the lateral section of the left third rib was resected, a video-assisted subtotal resection was performed with simultaneous reconstruction. The area of the post-resection sternocostal defect varied from 36 to 576 cm^2; the mean was 133 cm^2.

Autologous tissues and proprietary TiNi-based implants were used for chest wall reconstruction in all cases. Soft tissues were plastically reconstructed using a distant flap of the pectoralis major, the abdominal external oblique, or the latissimus dorsi. Depending on the size and location of the osteochondral defect indicated, thoracic cage anatomical rigidity was recovered by customized TiNi-based implants (Figure 1). A mesh depicted in Figure 1 is a low-profile implant made of a 100 μm superelastic TiNi wire using a knitting technology (Jersey knit). A double-layer mesh is the folded (two-ply) mesh applied in the case of a large excision defect. A strip is a flattened mesh sleeve (3.5 cm wide) implant made of a 100 μm superelastic TiNi wire using a circular knitting technology. A rib prosthesis (artificial rib) is a 6 mm thick customized sandwich consisting of a medullary wrought superelastic TiNi plate (2 mm thick) between cortical plates of porous SHS-TiNi secured together by a 150 μm superelastic TiNi wire wrapped around the device along its entire length.

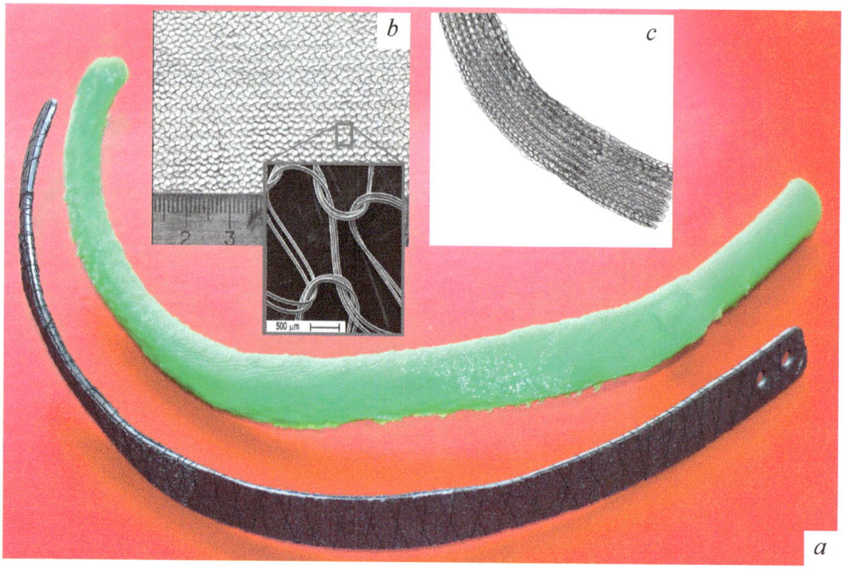

Figure 1. TiNi-based implants used in thoracic defect repairs: (**a**) customized artificial rib using a 3D-printed template, (**b**) knitted TiNi mesh, and (**c**) strip-flattened mesh sleeve.

In the early postoperative period, chest X-rays were taken to determine the position of the implants and identify any pathological changes. A chest CT with image reconstruction was performed three months post-surgery and at subsequent time points to evaluate the success or complications/aftereffects and to check the integrity of the thoracic cage. As a valid indicator of device consistency, functional outcomes, and patient status, we resorted to the Enneking modified scoring system (physical function, social role, pain, emotional acceptance, dexterity, etc.) [39].

3. Results and Discussion

Our surgical treatment of chest wall tumors accords well, in all cases, with an approach on the reconstruction procedure of a post-excision defect in the sternocostal framework reported in [1,3,6–13]. Artificial rib constructs for thoracic osteosynthesis prepared pursuant to a 3D model were a proper fit when the defect was being reconstructed and did not require any intraoperative correction (Figure 1), and thus the surgery time was shortened, as indicated in Table 1. Additionally, the latter provides an overview on the variants of chest wall resection, surgical/post-surgical features, and early complications determined case by case. The one-step surgery was more effective, and a good cosmetic effect was achieved. Options for the combined plasty of post-excision chest wall defects are summarized in Tables 3 and 4.

Table 3. Overview on variants of chest wall area resection, surgical/post-surgical features, and early complications.

No.	Defect Localization	No. of Excised Ribs	Lung Resection	Sternum Resection	Muscle Resection	Skin Resection	Others	Defect Area, cm^2	Operative Time, Min	ICU, Day	CTD, Day	PLoS, Day	Complications
1	Lat	2	CP	–	–	–	Pericardial	84	180	3	2	15	
2	Ant/Lat	2	LUL	–	PM	–	–	50	150	1	4	14	
3	Post/Lat	2	–	–	LD, PV	Yes	–	250	120	1	3	10	
4	Ant/Lat	4	WR	–	PM, Pm	–	–	198	185	1	5	14	
5	Ant	3	RUL	–	PM, Pm	–	–	78	140	2	4	12	
6	Ant	3	RUL	–	PM, Pm	Yes	–	98	240	5	8	18	pAL
7	Ant/Lat	1	–	–	Sc	–	–	36	80	–	1	7	
8	Ant/Lat	3	LUL	–	SM	–	–	98	210	1	5	15	
9	Post/Lat	2	RUL	–	PV	VB	–	84	180	5	4	21	pAF
10	Ant/Lat	4	–	–	PM	Yes	–	105	150	3	4	14	
11	Ant/Lat	1	–	–	–	–	–	70	85	0	1	5	
12	Ant/Lat	3	LUL	–	SM	–	–	92	130	2	6	12	PE, S
13	Ant/Lat	3	RUL	–	SM	–	–	98	110	2	5	12	
14	Ant/Lat	2	RUL	–	–	–	–	78	145	3	3	14	
15	Ant	8	–	Subtotal	PM	Yes	–	576	130	3	5	16	
Mean (min–max)	–	2.6 (1–8)	–	–	–	–	–	133 (36–576)	149 (80–240)	2.1 (0–5)	4 (1–8)	13.3 (5–21)	–

Note. Ant–anterior; CP–completion pneumonectomy; Lat–lateral; LD–*latissimus dorsi;* LUL–left upper lobectomy; pAF–paroxysm of atrial fibrillation; pAL–prolonged air leak; PE–pleural effusion; Pm–*pectoralis minor;* PM–*pectoralis major;* Post–posterior; PV–paravertebral muscles; RUL–right upper lobectomy; S–seroma; SM–serratus muscle; Sc–scalen muscle; VB–vertebral body; WR–wedge resection, ICU–intensive care unit, CTD-chest tube drain; PLoS–postoperative length of stay.

Table 4. Options used for combined plasty of chest wall defects.

No.	Mesh Type	Reinforcing Constituent Type	Soft Tissue Flap	Paradoxic Respiration
1	SL	PS	LT	–
2	SL	PS	Muscle	–
3	DL	PS	Muscle	–
4	DL	AR	Muscle	–
5	DL	2 AR	LT	–
6	DL	3 strips	Muscle	Yes
7	DL	–	Muscle	–
8	DL	AR	Muscle	–
9	DL	PS	LT	–
10	DL	3 AR	Muscle	–
11	DL	–	LT	–
12	DL	2 AR	LT	–
13	DL	2 AR	LT	–
14	DL	2 AR	LT	–
15	DL	3 AR	Muscle	–

Note. PS–pericostal sutures; SL–single-layer TiNi mesh; DL–double-layer TiNi mesh; AR–TiNi artificial rib, LT–local tissues.

Table 5. Summary of the surgical repair options in plasty of chest wall defects using TiNi-based implants.

Repair Option	No. of Patients	
	Abs.	%
Mesh + local tissues	1	6.7
Mesh + thoracodorsal flap	1	6.7
Double-layer mesh + local tissues	2	13.3
Double-layer mesh + pectoral flap	2	13.3
Double-layer mesh + strip (3 pcs) + pectoral flap	1	6.7
Double-layer mesh + rib prosthesis (1 pcs) + pectoral flap	2	13.3
Double-layer mesh + rib prosthesis (2 pcs) + local tissues	4	26.7
Double-layer mesh + rib prosthesis (3 pcs) + pectoral flap, external abdominal oblique muscle flap	2	13.3
Total	**15**	**100**

In eight patients with limited chest wall defects, mesh was used in combination with local tissues; a distant musculofascial flap was included as needed. In these cases, the mesh implant was draped and stitched circumferentially (along the perimeter of the defect) pursuant to the 'tent' method, having a tension that prevented pathological mobility of the repaired area. In three of eight cases, the area of the osteal chest lesion and thoracotomy coincided with the resection of lung tissue. Therefore, to rejoin the flawed ribs after thoracotomy, lobectomy, or pneumonectomy, a block pulley suture with polyester thread was used, and the threads in turn additionally reinforced the area of the reconstructed defect, as seen in Figure 2. In one patient with NSCLC, after extended combined upper right lobectomy with resection of ribs 2, 3, and 4, double-layer mesh with three strips placed between the folded layers was used. Extensive post-excisional chest wall lesions were repaired by combining double-layer mesh and artificial rib(s) placed in the surgical wound atop the double-layer mesh implant, complying with our original technique, as depicted in Figures 3 and 4.

The postsurgical period went smoothly for all of the patients. Patients were extubated in the operating room or within the first postoperative hours. No clinical signs of respiratory failure were observed. The patients became active within days of surgery. In the majority (93.3%) of cases, the surgical wound healed by first intention. In the intraoperative and postoperative periods, there were no lethal outcomes. Postoperative complications developed in three (20%) patients after simultaneous chest wall reconstruction and extended lobectomy for NSCLC. Analysis of postoperative complications indicated isolated cases of prolonged air ejection through the pleural drains, a paroxysm of atrial fibrillation, and pleuritis in combination with seroma of the subflap space, all of which were successfully managed conservatively.

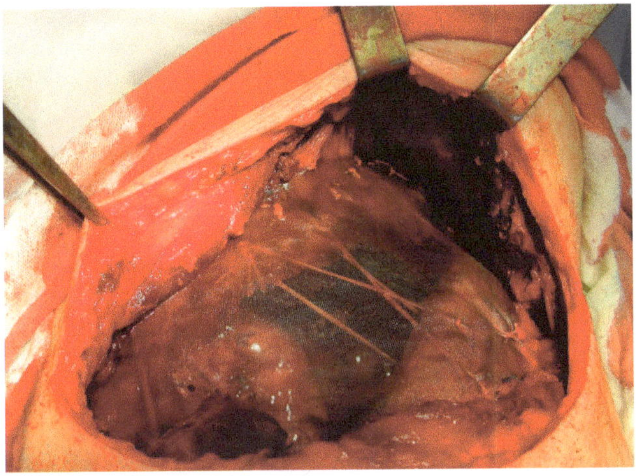

Figure 2. Combined left pneumonectomy followed by resection and repair of the chest wall. The ribs were fixed with pericostal sutures, and the thoracic wall defect was repaired using a knitted TiNi mesh implant.

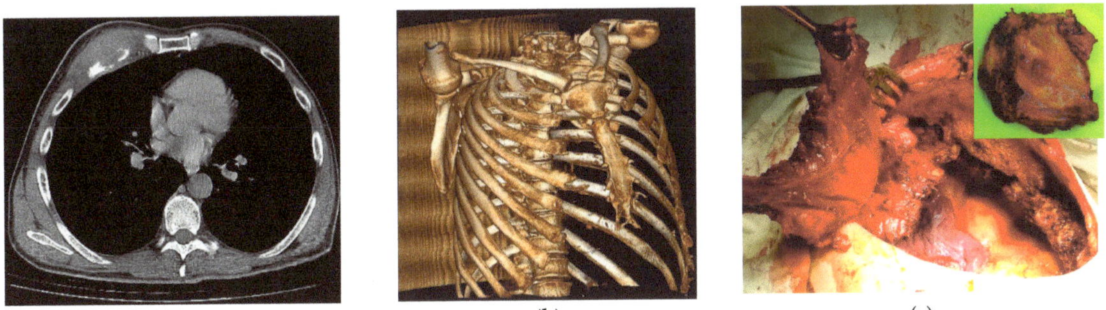

(a) (b) (c)

Figure 3. Patient XX, 48 y.o., with anterior chest wall metastatic mass invading ribs on the right from renal cell carcinoma: presurgical (**a**) axial CT thoracic scan and (**b**) 3D reconstruction of the chest osteochondral frame, (**c**) intraoperative view of the *pectoralis major/minor* to be excised together with anterior segments of three ribs (3rd–5th).

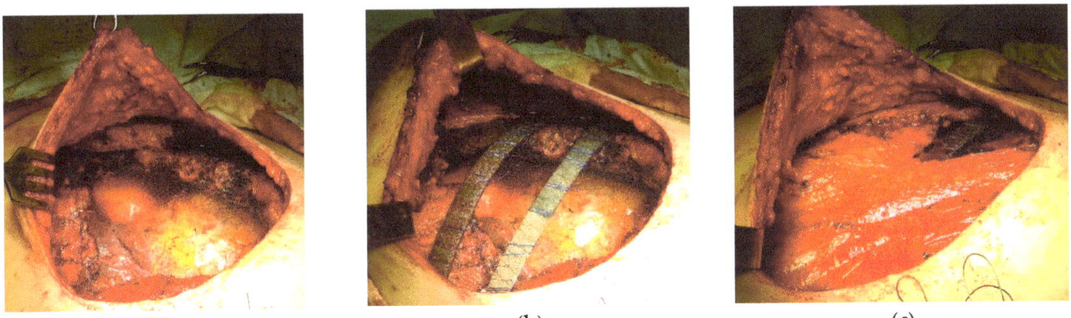

(a) (b) (c)

Figure 4. Patient XX, 48 y.o., intraoperative view of the large post-excision defect repaired by the double-layer TiNi knitted mesh implant stretched and sutured circumferentially (**a**), followed by the reinforcing artificial TiNi ribs is placed atop (**b**) and concealed with a prepared muscle flap (**c**), and the surgical wound is then draped with the cellulocutaneous flap.

In follow-up radiology examinations, no signs of movement of the artificial ribs relative to the initial position were detected in all cases. X-ray scans indicated that the implants were well incorporated into the host tissues, whereas thoracic cage distortion was negligible (Figure 5). In follow-up checks, and thus far, as a valid indicator of device consistency, functional outcomes, and patient status, we resorted to the Enneking modified scoring system (physical function, social role, pain, emotional acceptance, dexterity, etc.). On a five-point scale, with zero being the lowest, the functional result corresponded to excellent/good in 14 (93.3%) of the patients. The treatment outcome in the case of chest wall reconstruction with an area of 98 cm2 using TiNi knitted mesh and strips only was assessed as satisfactory. In this case, in the early post-surgical period, moderate pathological flotation of the plastically reconstructed chest area was noted that regressed in less than three months. An analysis of the given cases indicated that the defect could be considered yet sizeable to opt for at least one artificial rib that in combination with the knitted mesh would provide an optimal framework for the thoracic cage.

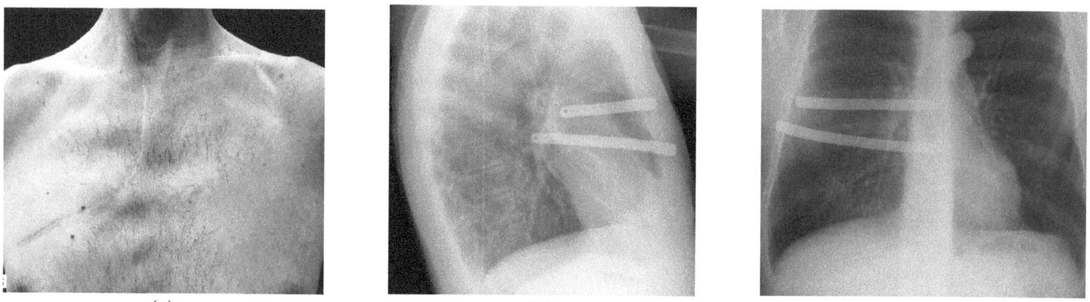

(a) (b) (c)

Figure 5. Patient XX, 48 y.o., intraoperative view of the large post-excision defect repaired by the double-layer TiNi knitted mesh implant stretched and sutured circumferentially (**a**), followed by the reinforcing artificial TiNi ribs is placed atop (**b**) and concealed with a prepared muscle flap (**c**), and the surgical wound is then draped with the cellulocutaneous flap.

We believe that in the reconstructive stage of eliminating a post-excisional thoracic defect, there is no need to take special steps to form the parietal pleura, and moreover this is not always possible when the tumor that was excised was malignant. In this case, the TiNi mesh implant, particularly in the double-layer format, deserves special attention as it successfully plays the pivotal role of a barrier membrane. The single-layer mesh implant is seen to be initially used (in two cases) because we did not have enough clinical experience yet, having no opportunity to evaluate all pros and cons. Further, to play it safe against possible risks and complications, we opted for the double-layer mesh. It is clear from general concepts that a thicker interface may contribute to the higher biointegration level in vivo, excluding the migration of customized artificial TiNi ribs with simultaneous maintenance of interlayer micro-motions of the composite structure. At the same time, it is necessary to maintain the chest framework while preserving the anatomical and physiological volume of the chest cavity in all cases, particularly when there are extensive thoracic defects. In limited defects, it is sufficient to use a mesh implant as the reinforcing element, and the implant can be supplemented with pericostal sutures. Extensive post-resection defects, however, require the artificial TiNi rib(s) as the reinforcing construct. Additionally, a multidisciplinary approach needs to include input from thoracic and plastic surgeons, as well as the clinical oncologist and anesthesiologist. Nevertheless, the number of rib prostheses in any given case is preoperatively chosen on an individual basis subject to the patient's anatomical features and the location of the defect. To restore the integumentary tissues of the chest, non-free skin/fatty tissue, skin/muscle, and muscle flaps, which have a good track record, should be used. Although the whole procedure was not performed following a standard, it addresses a prerequisite to improving surgical guidelines and for inclusion in the list of designated standard procedures.

4. Conclusions

Our experience indicates that the suggested surgical approach and tactics using one-step repair represent a promising technique even though the case is aggravated with extensive chest wall lesions, which can be performed safely and be recommended as a routine procedure with a high success rate. Combined TiNi-based implants seem to be very good reinforcing biomaterials that enable reliable repair of thoracic post-excisional defects of various sizes with good functional, clinical, and cosmetic outcomes.

Author Contributions: Conceptualization, E.T. and S.-B.K.; methodology, Y.Y.; validation, E.M. and A.O.; formal analysis, J.-H.K.; investigation, E.T., Y.Y. and T.C.; writing—original draft preparation, T.C. and E.T.; writing—review and editing, S.-B.K. and A.O.; visualization, Y.Y.; project administration, E.M.; funding acquisition, J.-H.K. All authors have read and agreed to the published version of the manuscript.

Funding: This work was supported by the Ministry of Education and Science of the Russian Federation, project No. 0721-2020-0022.

Institutional Review Board Statement: The study protocol was officially approved (approval code number—No 18/1152/2013, date 22 April 2013) by the Ethical Committee of the Siberian State Medical University.

Informed Consent Statement: Written informed consent to publish this work was obtained from all patients involved in the study.

Conflicts of Interest: The authors declare no conflict of interest. The authors have no financial interest in the products presented in this article.

References

1. David, E.A.; Marshall, M.B. Review of chest wall tumors: A diagnostic, therapeutic, and reconstructive challenge. *Semin. Plast. Surg.* **2011**, *25*, 16–24. [CrossRef]
2. Zarqane, H.; Viala, P.; Dallaudiere, B.; Vernhet, H.; Cyteva, C.; Larbi, A. Tumors of the rib. *Diagn. Interv. Imaging* **2013**, *94*, 1095–1108. [CrossRef] [PubMed]
3. Pfannschmidt, J.; Geisbusch, P.; Muley, T.; Hoffmann, H.; Dienemann, H. Surgical resection of secondary chest wall tumors. *Thorac. Cardiovasc. Surg.* **2005**, *53*, 234–239. [CrossRef] [PubMed]
4. Ferrigno, P.; Monaci, N.; Pangoni, A.; Comacchio, G.; Natale, G.; Faccioli, E.; Zuin, A.; Dell'Amore, A.; Rea, F. Extensive abdominal and chest wall resection and reconstruction for invasive squamous cell carcinoma of the skin. *J. Thorac. Dis.* **2020**, *12*, 45–49. [CrossRef] [PubMed]
5. Carter, B.; Gladish, G. MR imaging of chest wall tumors. *Magn. Reson. Imaging Clin. N. Am.* **2015**, *23*, 197–215. [CrossRef]
6. Kress, R.; Dalwadi, S.; Irani, A. R0 resection and reconstruction for a large, rapidly progressive chest wall sarcoma. *J. Cardiothorac. Surg.* **2018**, *13*, 127. [CrossRef] [PubMed]
7. Loi, M.; Mazzella, A.; Desideri, I.; Fournel, L.; Hamelin, E.; Icard, P.; Bobbio, A.; Alifano, M. Chest wall resection and reconstruction for lung cancer: Surgical techniques and example of integrated multimodality approach. *J. Thorac. Dis.* **2020**, *12*, 22–30. [CrossRef] [PubMed]
8. Merritt, R. Chest wall reconstruction without prosthetic material. *Thorac. Surg. Clin.* **2017**, *27*, 165–169. [CrossRef]
9. Marulli, G.; De Iaco, G.; Ferrigno, P.; De Palma, A.; Quercia, R.; Brascia, D.; Schiavon, M.; Mammana, M.; Rea, F. Sternochondral replacement: Use of cadaveric allograft for the reconstruction of ante-rior chest wall. *J. Thorac. Dis.* **2020**, *12*, 3–9. [CrossRef]
10. Sandri, A.; Donati, G.; Blanc, C.; Nigra, V.; Gagliasso, M.; Barmasse, R. Anterior chest wall resection and sternal body wedge for primary chest wall tumor: Recon-struction technique with biological meshes and titanium plates. *J. Thorac. Dis.* **2020**, *12*, 17–21. [CrossRef] [PubMed]
11. Sanna, S.; Brandolini, J.; Pardolesi, A.; Argnani, D.; Mengozzi, M.; Dell'Amore, A.; Solli, P. Materials and techniques in chest wall reconstruction: A review. *J. Vis. Surg.* **2017**, *3*, 95. [CrossRef] [PubMed]
12. Seder, C.; Rocco, G. Chest wall reconstruction after extended resection. *J. Thorac. Dis.* **2016**, *8*, S863–S871. [CrossRef] [PubMed]
13. Gao, E.; Li, Y.; Zhao, T.; Guo, X.; He, W.; Wu, W.; Zhao, Y.; Yang, Y. Reconstruction of anterior chest wall: A clinical analysis. *J. Cardiothorac. Surg.* **2018**, *13*, 124. [CrossRef] [PubMed]
14. Aranda, J.; Jimenez, M.; Rodriguez, M.; Varela, G. Tridimensional titanium-printed custom-made prosthesis for sternocostal re-construction. *Eur. J. Cardiothorac. Surg.* **2015**, *48*, e92–e94. [CrossRef]
15. Dzian, A.; Zivcak, J.; Penciak, R.; Hudak, R. Implantation of a 3D-printed titanium sternum in a patient with a sternal tumor. *World J. Surg. Oncol.* **2018**, *16*, 1315–1318. [CrossRef]

16. Wen, X.; Gao, S.; Feng, J.; Li, S.; Gao, R.; Zhang, G. Chest-wall reconstruction with a customized titanium-alloy prosthesis fabricated by 3D printing and rapid prototyping. *J. Cardiothorac. Surg.* **2018**, *13*, 4. [CrossRef]
17. Zhang, H.; Zhao, J.; Li, X.; Huang, L.; Wang, L. Necessity of pleura repair in the chest wall reconstruction with three-dimensional printed tita-nium implant. *J. Thorac. Dis.* **2020**, *12*, 2713–2716. [CrossRef]
18. Weyant, M.; Bains, M.; Venkatraman, E.; Downey, R.; Park, B.; Flores, R.; Rizk, N.; Rusch, V. Results of chest wall resection and reconstruction with and without rigid prosthesis. *Ann. Thorac. Surg.* **2006**, *81*, 279–285. [CrossRef]
19. Bille, A.; Okiror, L.; Karenovics, W.; Routledge, T. Experience with titanium devices for rib fixation and coverage of chest wall defects. *Interact. Cardiovasc. Thorac. Surg.* **2012**, *15*, 588–595. [CrossRef]
20. Berthet, J.; Canaud, L.; D'Annoville, T.; Alric, P.; Marty-Ane, C. Titanium plates and Dualmesh: A modern combination for reconstructing very large chest wall defects. *Ann. Thorac. Surg.* **2011**, *91*, 1709–1716. [CrossRef]
21. Ong, K.; Ong, C.; Chua, Y.; Fazuludeen, A.; Ahmed, A. The painless combination of anatomically contoured titanium plates and porcine dermal collagen patch for chest wall reconstruction. *J. Thorac. Dis.* **2018**, *10*, 2890–2897. [CrossRef] [PubMed]
22. Lu, X.; Xiong, S.; Chen, Y.; Zhao, F.; Hu, Y.; Guo, Y.; Wu, B.; Huang, P.; Yang, B. Effects of statherin on the biological properties of titanium metals subjected to different surface modification. *Colloids Surf. B Biointerfaces* **2020**, *188*, 110783. [CrossRef] [PubMed]
23. Van Oirschot, B.; Meijer, G.; Bronkhorst, E.; Narhi, T.; Jansen, J.; Van den Beucken, J. Comparison of different surface modifications for titanium implants installed into the goat iliac crest. *Clin. Oral Implant. Res.* **2014**, *27*, e57–e67. [CrossRef] [PubMed]
24. Muhamedov, M.; Kulbakin, D.; Gunther, V.; Choynzonov, E.; Chekalkin, T.; Hodorenko, V. Sparing surgery with the use of TiNi-based endografts in larynx cancer pa-tients. *J. Surg. Oncol.* **2015**, *111*, 231–236. [CrossRef] [PubMed]
25. Shtin, V.; Novikov, V.; Chekalkin, T.; Gunther, V.; Marchenko, E.; Choynzonov, E.; Kang, S.B.; Chang, M.J.; Kang, J.H.; Obrosov, A. Repair of orbital post-traumatic wall defects by custom-made TiNi mesh endografts. *J. Funct. Biomater.* **2019**, *10*, 27. [CrossRef] [PubMed]
26. Chernyshova, A.; Kolomiets, L.; Chekalkin, T.; Chernov, V.; Sinilkin, I.; Gunther, V.; Marchenko, E.; Baigonakova, G.; Kang, J.H. Fertility-sparing surgery using knitted TiNi mesh implants and sentinel lymph nodes: A 10-year experience. *J. Investig. Surg.* **2021**, *34*, 1110–1118. [CrossRef] [PubMed]
27. Kulbakin, D.; Chekalkin, T.; Muhamedov, M.; Choynzonov, E.; Kang, J.H.; Kang, S.B.; Gunther, V. Sparing surgery for the successful treatment of thyroid papillary carcinoma invading the trachea: A case report. *Case Rep. Oncol.* **2016**, *9*, 772–780. [CrossRef]
28. Gunther, V.; Radkevich, A.; Kang, S.B.; Chekalkin, T.; Marchenko, E.; Gunther, S.; Pulikov, A.; Sinuk, I.; Kaunietis, S.; Podgorniy, V.; et al. Study of the knitted TiNi mesh graft in a rabbit cranioplasty model. *Biomed. Phys. Eng. Express* **2019**, *5*, 027005. [CrossRef]
29. Yasenchuk, Y.; Marchenko, E.; Gunther, V.; Radkevich, A.; Kokorev, O.; Gunther, S.; Baigonakova, G.; Hodorenko, V.; Chekalkin, T.; Kang, J.H.; et al. Biocompatibility and clinical application of porous TiNi alloys made by self-propagating high-temperature synthesis (SHS). *Materials* **2019**, *12*, 2405. [CrossRef]
30. Yasenchuk, Y.; Marchenko, E.; Baigonakova, G.; Gunther, S.; Kokorev, O.; Gunter, V.; Chekalkin, T.; Topolnitskiy, E.; Obrosov, A.; Kang, J.H. Study on tensile, bending, fatigue, and in vivo behavior of porous SHS-TiNi alloy used as a bone substitute. *Biomed. Mater.* **2021**, *16*, 021001. [CrossRef]
31. Gunther, V.; Yasenchuk, Y.; Chekalkin, T.; Marchenko, E.; Gunther, S.; Baigonakova, G.; Hodorenko, V.; Kang, J.H.; Weiss, S.; Obrosov, A. Formation of pores and amorphous-nanocrystalline phases in porous TiNi alloys made by self-propagating high-temperature synthesis (SHS). *Adv. Powder Technol.* **2019**, *30*, 673–680. [CrossRef]
32. Yasenchuk, Y.; Gunther, V.; Marchenko, E.; Chekalkin, T.; Baigonakova, G.; Hodorenko, V.; Gunther, S.; Kang, J.H.; Weiss, S.; Obrosov, A. Formation of mineral phases in self-propagating high-temperature synthesis (SHS) of porous TiNi alloy. *Mater. Res. Express* **2019**, *6*, 056522. [CrossRef]
33. Kokorev, O.; Hodorenko, V.; Chekalkin, T.; Gunther, V.; Kang, S.B.; Chang, M.J.; Kang, J.H. Evaluation of allogenic hepato-tissue engineered in porous TiNi-based scaffolds for liver regeneration in a CCl4-induced cirrhosis rat model. *Biomed. Phys. Eng. Express* **2019**, *5*, 025018. [CrossRef]
34. Kokorev, O.; Chekalkin, T.; Marchenko, E.; Yasenchuk, Y.; Gunther, S.; Serebrov, V.; Chernyshova, A.; Obrosov, A.; Uludintceva, E.; Kang, J.H. Exploring the role of surface modifications of TiNi-based alloys in evaluating in vitro cytocompatibility: A comparative study. *Surf. Topogr. Metrol. Prop.* **2020**, *8*, 045015. [CrossRef]
35. Aihara, H.; Zider, J.; Fanton, G.; Duerig, T. Combustion synthesis porous Nitinol for biomedical applications. *Int. J. Biomater.* **2019**, *2019*, 4307461. [CrossRef]
36. Topolnitskiy, E.; Dambayev, G.; Gyunter, V. The replacement of postresectional defects of the thorax with the use of tissue im-plant of nanostructural nickelid-titan thread. *Khirurgiia* **2011**, *10*, 47–53.
37. Zheravin, A.; Gyunter, V.; Anisenya, I.; Garbukov, E.; Zhamgaryan, G.; Bogoutdinova, A. Reconstruction of the chest wall using titanium-nickelid for cancer patients. *Sib. J. Oncol.* **2015**, *3*, 31–38.
38. Kudrjavtsev, A.; Zheravin, A.; Anikeeva, O.; Polovnikov, E.; Yarmoshuk, S.; Drobyazgin, E. Treatment of a patient with sarcoma of sternum, ribs and invasion into peri-card and pleura. *Patologiya Krovoobrashcheniya i Kardiokhirurgiya* **2015**, *1924*, 124–129.
39. Wada, T.; Kawai, A.; Ihara, K.; Sasaki, M.; Sonoda, T.; Imaeda, T.; Yamashita, T. Construct validity of the Enneking score for measuring function in patients with malignant or aggressive benign tumours of the upper limb. *J. Bone Joint Surg.* **2007**, *89*, 659–663. [CrossRef]

Journal of
Functional Biomaterials

MDPI

Article

Molar Septum Expansion with Osseodensification for Immediate Implant Placement, Retrospective Multicenter Study with Up-to-5-Year Follow-Up, Introducing a New Molar Socket Classification

Samvel Bleyan [1,†], João Gaspar [2,*,†], Salah Huwais [3], Charles Schwimer [3,4], Ziv Mazor [5,6], José João Mendes [2] and Rodrigo Neiva [3]

1 Private Practice, 3/1 Tatarskaya St., 115184 Moscow, Russia; samvel32@gmail.com
2 Clinical Research Unit (CRU), Centro de Investigação Interdisciplinar Egas Moniz (CiiEM), Egas Moniz Cooperativa de Ensino Superior, 2829-511 Almada, Portugal; jmendes@egasmoniz.edu.pt
3 Department of Periodontics, School of Dental Medicine, University of Pennsylvania, Philadelphia, PA 19104, USA; shuwais@versah.com (S.H.); f40chuck@gmail.com (C.S.); rneiva@upenn.edu (R.N.)
4 Department of Periodontics, School of Dental Medicine, University of Pittsburgh, Pittsburgh, PA 15260, USA
5 Private Practice, 144 Begin St., Tel Aviv 6492102, Israel; mazor2@yahoo.com
6 Department of Periodontics, Titu Maiorescu University, 040441 Bucharest, Romania
* Correspondence: jgaspar@egasmoniz.edu.pt
† S.B. and J.G. contributed equally to this work and are equally first authors.

Abstract: The ideal positioning of immediate implants in molar extraction sockets often requires the osteotomy to be in the interradicular septum, which can be challenging in some cases, with traditional site preparation techniques. Patients who had undergone molar tooth extraction and immediate implant placement at five different centers, and followed up between August 2015 and September 2020, were evaluated. Inclusion criteria were use of the osseodensification technique for implant site preparation. The primary outcome was septum width measurement pre-instrumentation and osteotomy diameter post expansion. Clinical outcomes, such as implant insertion torque (ISQ) and implant survival rate, were also collected. A total of 131 patients, who received 145 immediate implants, were included. The mean overall septum width at baseline was 3.3 mm and the mean osteotomy diameter post instrumentation was 4.65 mm. A total of ten implants failed: seven within the healing period and three after loading; resulting in a cumulative implant survival rate of 93.1%. This retrospective study showed that osseodensification is a predictable method for immediate implant placement with interradicular septum expansion in molar extraction sockets. Furthermore, it allowed the introduction of a new molar socket classification. In the future, well-designed controlled clinical studies are needed to confirm these results and further explore the potential advantages of this technique.

Keywords: osseodensification; immediate implant placement; septum expansion; osteotomy; osseointegration

1. Introduction

Immediate implant placement (IIP) into fresh extraction sockets has aroused interest since it was initially described [1] and has been considered a predictable therapeutic approach for both anterior and posterior sites, with survival rates comparable to implants placed in healed ridges [2–5]. An 11-year retrospective study of 300 implants immediately placed in molar extraction sockets reported an overall survival rate of 97.3% [6]. Furthermore, a systematic review [7] of outcomes following immediate molar implant placement demonstrated a survival rate of 98%, with no significant differences between maxilla and mandible. More recently, another systematic review and meta-analysis [8] of immediate implants in molar extraction sites demonstrated success rates of 93.3% after 1 year of follow-up.

This treatment alternative offers several advantages in comparison to the classic delayed approach, namely a single surgical intervention, with a reduction in overall treatment time and, therefore, increased patient satisfaction [9,10]. However, its success was reported to be influenced by several factors, including the need for atraumatic extraction to preserve favorable socket anatomy, as well as the effect of site instrumentation to achieve an adequate initial implant stability [9]. Implant primary stability and adequate insertion torque are considered critical aspects for successful IIP [5,6,11,12]. Several challenges have been described in achieving initial stabilization in molar extraction sockets. These include the width of the extraction socket, poor bone quality, inadequate interradicular bone septum width, and anatomical limitations beyond the apex of the roots, such as the inferior alveolar canal in the mandible or the maxillary sinus in the maxilla [13]. Thus, flapless tooth extraction with minimal trauma and gentle separation of the roots is essential to preserve a favorable anatomy and to allow the placement of the implant within the socket itself, when needed [10,14]. In addition, implant primary stability and insertion torque is related to the density of the bone pre and post site preparation. Bone density is known to have a direct effect on implant stability, as the denser the bone surrounding the osteotomy walls, the higher the insertion torque and the ISQ values [15]. Both these parameters are influenced by the drilling protocol [16,17], so enhancing the bone density during osteotomy preparation may improve clinical success, especially in the maxilla, due to its typically lower bone density compared to the mandible [18].

Smith and Tarnow [14] classified molar sockets based on the amount of interradicular septal bone in relation to implant placement into three types: Type A sockets have sufficient septal bone bulk to circumferentially contain the implant. Type B sockets have enough septal bone bulk to stabilize the implant, but not fully surround it. On the other hand, Type C sockets have insufficient septal bone to stabilize the implant without engaging the socket walls, so this would either indicate the placement of ultra-wide diameter implants or a delayed placement approach. According to several authors [19–21], immediate implant placement in molar extraction sockets using ultra-wide implants demonstrates a predictable outcome, with reduced bone loss and stable soft and hard tissue conditions. However, in a systematic review [7] conducted in 2016, ultra-wide implants (>6–9 mm) were found to have a significantly higher failure rate than implants of 4 to 6 mm diameter. More recently, Ragucci et al. [8] also recommended the use of implants of <5 mm diameter for immediate placement in molar extraction sockets. Therefore, implant placement in the interradicular septum is usually considered the best option for an immediate molar implant, not only in terms of correct 3D positioning, but also regarding implant survival [10].

Recently, a novel non-subtractive surgical technique for implant site preparation termed osseodensification (OD) was introduced [22]. Contrary to traditional extractive drilling protocols, it preserves bone and enhances its plasticity, utilizing specially designed burs that rotate in a non-cutting (counter-clockwise) direction to gradually expand the osteotomy, while simultaneously compacting bone into its trabecular spaces, increasing the density of the site [22–25]. Furthermore, OD was shown to enhance implant primary stability, due to the compaction auto-grafting and the associated spring-back effect [22,26]; increasing bone-to-implant contact (BIC) upon implant placement [24,25]. These autografted bone particles in the trabecular spaces act as nucleation for faster bone formation around the implant, potentially shortening the healing time [23–25,27]. Large-animal histological studies have demonstrated that this high stability at the day of surgery is maintained throughout the implant healing process, regardless of the implant macro- or micro-geometry [24,25,27]. In a recent multicenter controlled clinical trial, OD also demonstrated significantly higher insertion torque and ISQ values compared to conventional subtractive drilling for all implant dimensions, with the exception of short implants, regardless of the jaw and area operated, and irrespective of the evaluation period [28]. Osseodensification's ability to plastically expand trabecular bone with compaction autografting, to facilitate implant placement with sufficient stability and adequate healing in sites with less than optimum bone quantity and quality, was documented in both in vivo and clinical data [23,29]. Trisi et al. [23] was able to demonstrate, in a large animal histological study, the predictability of placing a 5-mm implant in 5-mm wide ridge in sheep iliac crest with adequate healing. Koutouzis and Huwais [29] confirmed his findings in a clinical controlled study that demonstrated a 93% success rate for 38 implants placed in plastically expanded alveolar ridges via osseodensification in 21 patients. In addition to ridge plastic expansion, osseodensification has also been reported to enhance dental implant's short and long-term success rate, regardless of their macro- or micro-geometry, in several clinical scenarios, including immediate loading [30–32], as well as to facilitate implant placement in conjunction with crestal sinus graft, with a high success rate [33–35].

The aim of the present multicenter retrospective study was to assess the effectiveness of interradicular septum expansion with osseodensification site preparation for immediate implant placement in molar extraction sockets.

2. Materials and Methods

This retrospective analysis followed the World Medical Association Declaration of Helsinki and the directives given by the Egas Moniz Ethics Commission (CEEM) at Egas Moniz Cooperativa de Ensino Superior, Monte de Caparica, Portugal, which does not require ethical approval for retrospective clinical studies.

An informed consent form was signed by all patients included in the study, both for the clinical procedure and follow-up appointments. All treatment steps and data collection were part of the routine procedures at the centers, and no extra measures were taken for the purpose of the study. All examiners were blind, since a random case number was allocated to the extracted data, ensuring patient anonymity and data protection. The study was structured following the STROBE statement [36].

2.1. Selection Criteria and Surgical Technique

Patients who had undergone molar tooth extraction and immediate implant placement with osseodensification at five different centers (S.B., J.G., S.H., C.S., Z.M.), followed up between August 2015 and September 2020, were evaluated. Inclusion criteria included patients with molar extraction sockets that had an interradicular septum of at least 2.5 mm width, use of the osseodensification technique for implant site preparation, and follow-up of a minimum of 12 months after loading with a definitive implant-supported restoration. Exclusion criteria comprised an initial septum width <2.5 mm, history of radiotherapy, bisphosphonate medication, active periodontal disease, uncontrolled diabetes, heavy smoking (>20 cigarettes/day), and local acute apical abscess. All patients had a cone beam computed tomography (CBCT) prior to surgical procedure.

All interventions were performed by experienced surgeons, who followed standardized surgical technique. After local anesthesia with articaine (4%) and epinephrine (1:200,000), flapless tooth extraction, as atraumatic as possible, was performed after separation of the roots with a long thin diamond bur, in order to preserve the interradicular bone and the general socket anatomy. The socket was then thoroughly curetted to detach any granulation tissue that could potentially impair healing.

Septum width was directly measured post molar extraction. Measurement was recorded at the narrowest width of the septum. Implant site preparation started with a pilot drill, in clockwise motion, in the center of the septum, until 1 mm deeper than the planned implant length. Densah® Burs (Versah, LLC, Jackson, MI, USA) were then sequentially used in OD mode (counterclockwise, drilling speed 800–1500 rpm, with copious irrigation) in small increments to gradually expand the osteotomy, until reaching the desired width for the planned implant diameter (Figure 1).

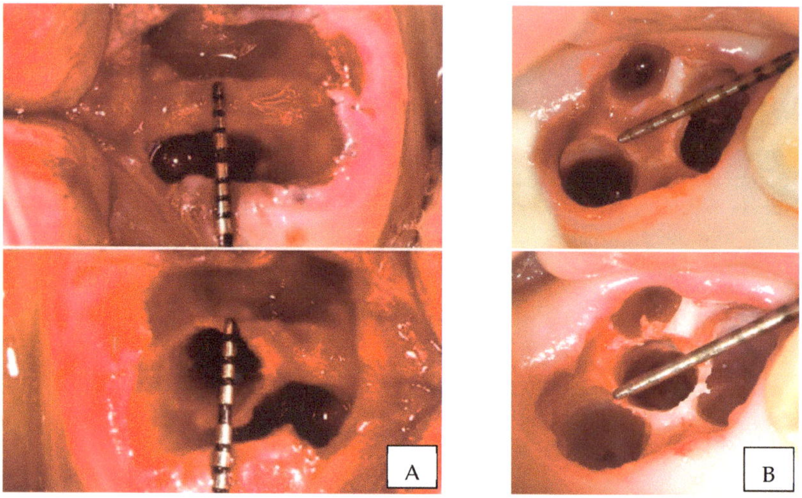

Figure 1. Clinical examples of interradicular septum expansion after implant site preparation with osseodensification ((**A**). Mandibular first molar; (**B**). Maxillary second molar).

Osteotomy diameter as a reflection of septum width expansion was then directly measured and recorded after site instrumentation. Although each center used the implant company of their choice, all implants placed were conical, bone-level, and with internal connection (Table 1). After implant placement at the adequate depth, the gaps were filled with allograft or alloplastic, depending on each center's preference and either a customized or a large stock sealing healing abutment was placed, with no attempt to coronally advance the flaps for primary intention healing. The insertion torque value was registered, and implant stability was measured using resonance frequency analysis, immediately after implant insertion (primary stability) and after healing, before final impression (secondary stability).

Table 1. Overview of implants included in the retrospective analysis.

Implant Company	Number of Implants Placed	Number of Implants Failed
Dentium	35	1
Adin	35	3
Megagen	26	1
Neobiotech	21	2
Zimmer	14	3
Paltop	6	0
IDI	5	0
Nobel Biocare	3	0
Total	145	10

The osseointegration period varied according to the decision of each clinician, based on the records mentioned above and on bone quantity and quality, with a minimum of 3 months. Despite not following a standardized medication protocol, all patients were prescribed post-operative antibiotics for 7–10 days, based on each center's preference.

2.2. Variables and Statistics

Data regarding patient characteristics (age and gender); tooth location; date of surgical and restorative procedure; septum width before and after site preparation and expansion; insertion torque and ISQ at baseline and after osseointegration; implant width and length; time of loading; osseointegration success rate; and final follow-up appointment were collected from the patients' clinical files. The primary outcome was septum width measurement pre-instrumentation and osteotomy diameter post expansion. Descriptive statistics were conducted using IBM® SPSS® Statistics software (SPSS for Mac, Version 26.0. SPSS Inc. Chicago, IL, USA). A Kaplan–Meier curve was used to analyze the survival rate of implants placed. This curve was adjusted to 12 months, because it was the minimum follow-up common to all implants.

3. Results

A total of 131 patients, 90 women and 41 men, with a mean age of 52 years (range 27–80), who received 145 immediate implants in molar extraction sockets, were included (Figure 2). The mean follow-up of the included patients was 36 months (range 12–60 months). Reasons for tooth extraction were endodontic treatment failure, root fracture, or non-restorable teeth. No extracted teeth sockets for periodontal reasons were included.

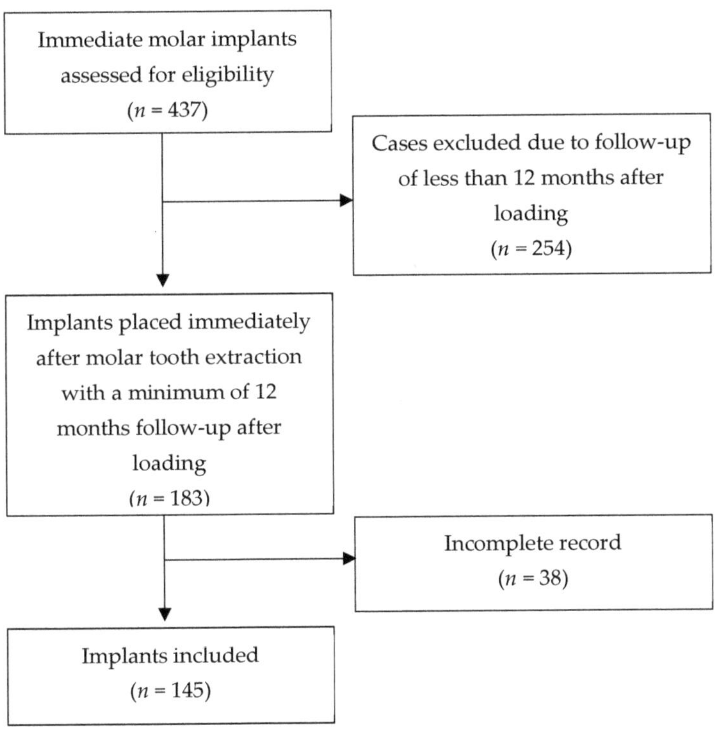

Figure 2. Flowchart representative of implants included in the retrospective analysis.

A total of 87 implants were placed in the mandible (72 in first molar sites and 15 in second molar sites) and 58 in the maxilla (53 in first molar sites and 5 in second molar sites), as shown in Figure 3. Maxillary sockets had higher mean values of interradicular septum width compared to those in the mandibular, as described in Figure 4.

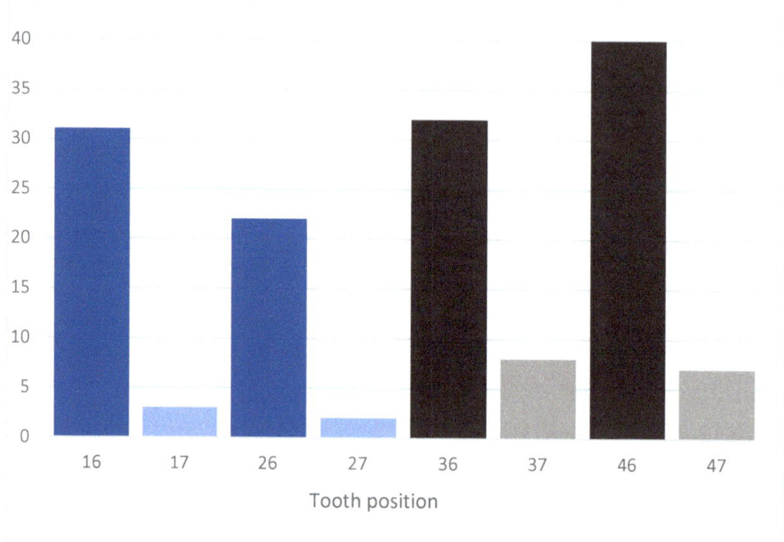

Figure 3. Number of implants placed, according to tooth position.

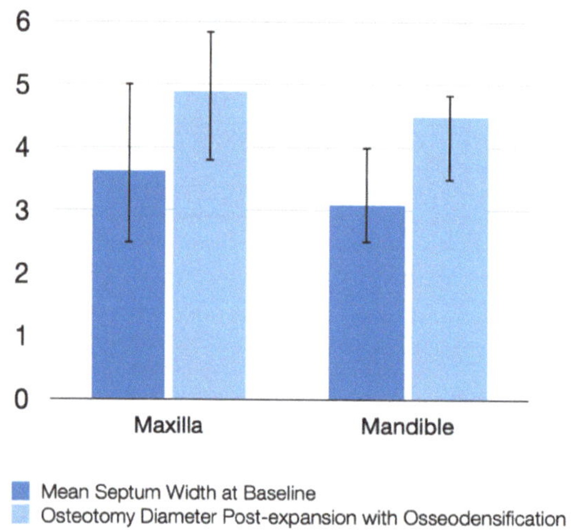

Figure 4. Mean septum width before instrumentation and osteotomy diameter post-expansion with osseodensification.

The mean overall septum width at baseline was 3.3 mm, and the mean osteotomy diameter post instrumentation was 4.65 mm after expansion with osseodensification (Figure 5).

Implant stability was measured by both insertion torque (ITV) and ISQ values. ITV was higher in the mandible (mean 46.72 N cm; range 30–60 N cm) than in the maxilla (mean 41.12 N cm; range 20–60 N cm), with an overall mean value of 44.48 ± 8.2 N cm (Figure 6).

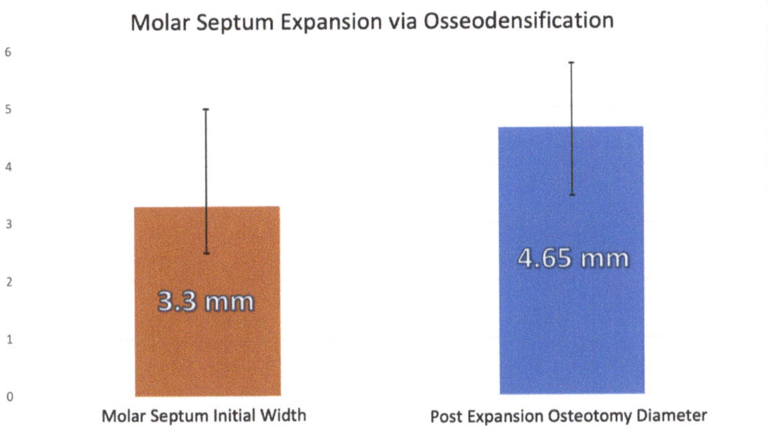

Figure 5. Mean overall septum width at baseline and osteotomy diameter post-expansion with osseodensification.

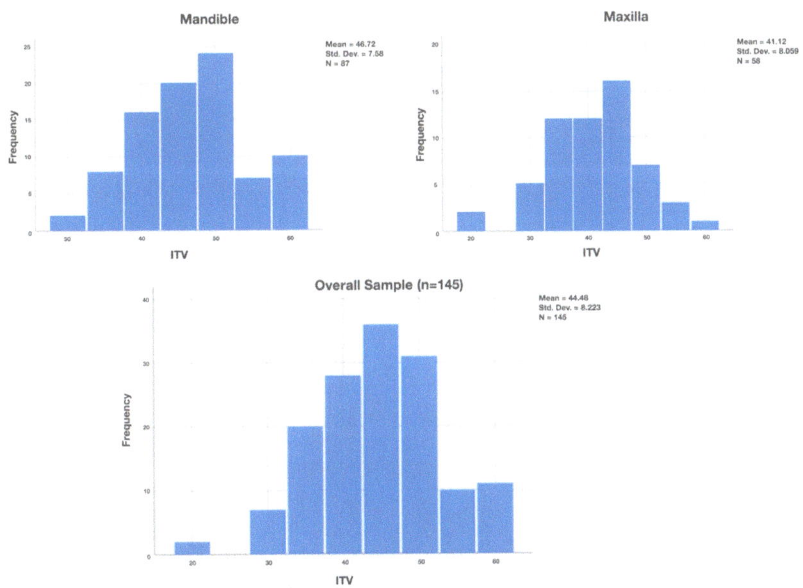

Figure 6. Insertion torque value (ITV) of implants placed.

Only 6.2% of the implants had an ITV <3 5 N cm, while 35.9% had an ITV $\geq$ 50 N cm. Mean ISQ was 72.8 (range 60–82) at baseline on the day of surgery (ISQS) and 78.9 (range 70–88) after the osseointegration period, before final impression (ISQR), as described in Table 2. Implant diameter ranged from 4.2 to 6.4 mm, and length ranged from 10 to 13 mm, depending on the implant system used in each center. A total of ten implants (four in the mandible and six in the maxilla) failed (Table 3): seven within the healing period before final impression and three after loading, resulting in a survival rate of 93.1%. Only two centers included smoker patients (n = 6), who did not experience implant failure; therefore, no correlation could be assessed between smoking and implant failure. The Kaplan–Meier estimator predicted a 93.1% survival rate at 12 months follow-up (Figure 7).

Table 2. Mean ISQ measurement on day of surgery and in restorative phase after osseointegration period.

		ISQS	ISQR
Maxilla	Mean	71.47	77.26
	N	58	54
	Std. Deviation	4.231	3.004
Mandible	Mean	73.72	79.88
	N	87	84
	Std. Deviation	4.358	3.730
Total	Mean	72.82	78.86
	N	145	138
	Std. Deviation	4.434	3.684

ISQS–ISQ in day of surgery; ISQR–ISQ in restorative phase.

Table 3. Description of failed implants.

Implant Company	Diameter	ITV	ISQ	Septum Pre	Septum Post	Timing of Failure
Neobiotech	5	55	76	3.5	4.8	After
Neobiotech	5	35	65	3.4	4.8	Before
Dentium	5	40	68	2.8	4.5	Before
Zimmer	5.2	20	63	5	5.5	Before
Zimmer	5	20	62	4	5.5	Before
Zimmer	4.7	30	60	2.5	4.5	After
Megagen	5.0	30	70	3.5	4.8	After
Adin	4.3	50	75	3	4.5	Before
Adin	5	45	70	3	4.5	Before
Adin	4.3	50	74	2.5	3.8	Before
Mean	4.85	37.5	68.3	3.32	4.7	

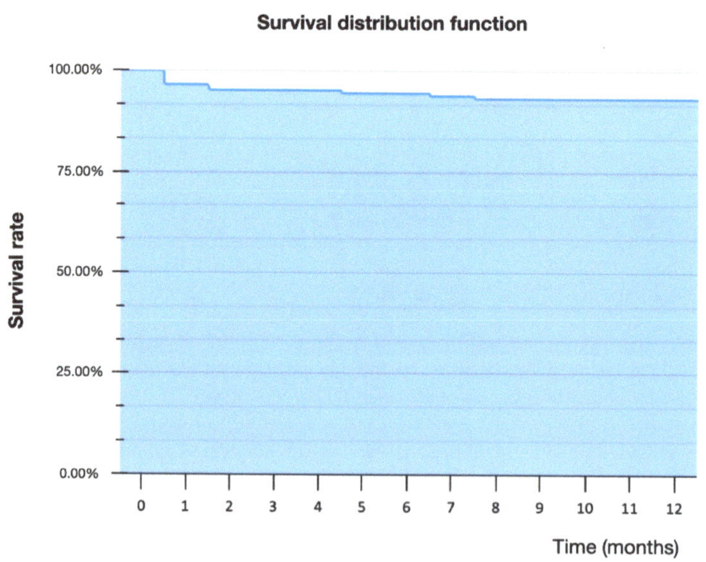

Figure 7. Kaplan–Meier survival curve for survival estimate.

4. Discussion

According to a recent systematic review and meta-analysis [8], the suggested approach for IIP in molar extraction sockets includes a flapless procedure, a one-stage implant placement, grafting the gap, and the use of implants with <5 mm diameter. Flapless surgery may not only contribute to decreased operative time, but also to faster healing, reduction of peri-implant tissue collapse, less postoperative complications, and improved patient comfort [8]. This approach was followed by all centers in this study, except for the implant diameter. In fact, 64.2% of the implants placed by all centers had a diameter $\geq$ 5 mm. The mean overall septum width at baseline was 3.3 mm and the mean implant diameter of all implants placed was 4.96 mm, which demonstrates the potential of the osseodensification technique to preserve the bony housing and expand the septum; thus, allowing predictably placing wider diameter implants compared to the conventional osteotomy technique.

Walker et al. [11] assessed the relationship between insertion torque values and clinical outcomes and reported an implant survival rate in immediate molar implants of 86%, when insertion torque was low, and 90% to 96% when IT was medium to high, respectively. This tendency was also observed in our study, since four out of the nine implants that had an insertion torque < 35 N cm failed. Moreover, the mean insertion torque of the implants that failed (n = 10) was 37.5, while the mean insertion torque of the successful implants (n = 135) was 45. Regarding ISQ, implants that ended up failing had lower mean ISQ values (68.3) on the day of surgery, compared to implants successfully integrated and loaded (73.2). In a recently published multicenter controlled clinical trial [28], OD drilling demonstrated significantly higher IT and temporal ISQ values relative to more conventional subtractive drilling techniques for all implant dimensions, with the exception of short implants. Therefore, we may assume that the implant survival would probably be lower with a traditional drilling protocol.

All sockets evaluated in this retrospective analysis were grafted with either allograft or alloplast (Novabone®). Bone grafting of the remaining socket voids adjacent to an immediate implant is not essential for osseointegration to occur, especially if the outer walls of the socket are intact [10,37,38]. However, its combination with a customized healing abutment, acting as a prosthetic socket seal device minimizes the amount of ridge contour change after tooth extraction and IIP, thereby contributing to better esthetics and restorative contour [10,38], as observed in this study (Figure 8).

Pre-operative CBCT is an essential and effective diagnostic method to evaluate socket anatomy and to define the most suitable treatment approach for each case, as well as minimizing the risk of damaging vital structures [39]. Historically, a minimum 3 mm width of interradicular septal bone (ISB) was deemed important to achieve initial stabilization of an immediate molar implant [40]. In this study, twenty-three extraction sockets had an ISB width of 2.5 mm and one had 2.8 mm. Nevertheless, the osseodensification technique used for implant site preparation allowed adequate septum expansion of all these sites, to create osteotomies diameters in a range of 3.5–4.5 mm, thereby providing adequate implant stability upon insertion. Moreover, since it pushes bone in a both lateral and apical direction, instead of removing it, osseodensification also predictably allows sinus elevation using a crestal approach in maxillary molar sockets with reduced residual bone height below the sinus floor [33].

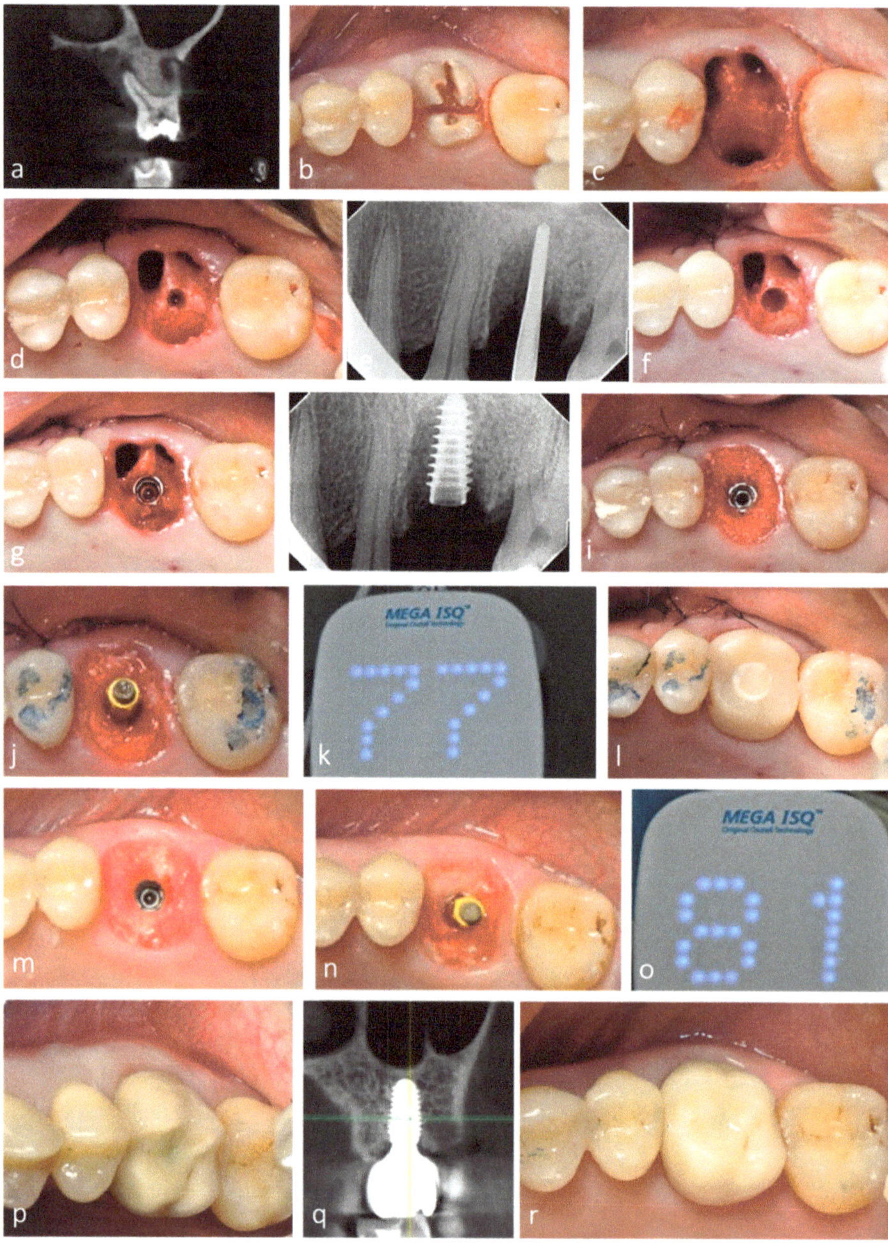

Figure 8. Representative clinical case with 3-year clinical and radiographic follow-up. (**a**) CBCT of maxillary left first molar showing periapical infection with extensive bone loss buccally and palatally. (**b**) Root section for tooth extraction as atraumatic as possible. (**c**) Septum preservation after extraction. (**d**,**e**) Initial osteotomy depth at 10 mm. (**f**) Implant site preparation, optimized with osseodensification. (**g**,**h**) Implant placed in the expanded septum. (**i**) Allograft placed in the root sockets to fill the extraction socket. (**j**–**l**) Adequate implant stability allowed for the placement of a fully contoured customized socket sealing healing abutment out of occlusion. (**m**) Healing after 3 months, with contour maintenance. (**n**,**o**) ISQ measurement after osseointegration period. (**p**–**r**) Clinical and radiographic follow-up after 3 years.

Traditionally, Smith and Tarnow classification type B sockets [14] with narrow septa are commonly managed by clinicians through the placement of an immediate implant into the palatal root socket of maxillary molars or into one of the two mandibular molar root sockets [41]. However, this may lead to potential food impaction and tissue inflammation, due to poor emergence profile of the restoration. Furthermore, Smith et al. [42] observed, in a retrospective study of 300 implants, that there is a direct correlation between the horizontal implant–tooth distance and the incidence of adjacent tooth decay. Therefore, immediate implant placement in the mesial or distal molar root sockets may significantly increase the risk of decay in the furthest tooth. Accordingly, the ideal implant positioning in molar sockets will most often require the osteotomy to be in the septum. Osseodensification may facilitate the preservation and the expansion of the interradicular septum, thus enhancing the ability to predictably place implants with adequate stability in both type B and C sockets, as shown in this study (Figure 9).

The traditional classification of molar extraction sockets by Smith and Tarnow [14] is based on the amount of interradicular septal bone remaining post instrumentation around immediately placed implants, but it does not take into account the specific measurement of the septum width pre-instrumentation nor pre-implant placement. Furthermore, the specific type of the socket in this classification is dependent on the diameter of the implant placed. Therefore, the authors of the present study propose a new diagnostic classification that is based on the initial septal bone width prior to site preparation and implant placement, which would allow adequate treatment planning. The new classification (Figure 10) includes four categories: S-I: septum initial width >4 mm; S-II: septum initial width = 3–4 mm; S-III: septum initial width = 2–3 mm; and S-IV: septum initial width < 2 mm/no septal bone. The relevance of this new diagnostic classification is related to the fact that, with osseodensification instrumentation, and due to bone preservation and plastic expansion, it is possible to convert type B sockets into type A, and type C into type B [14], as was observed in this study. According to our classification, only S-IV sockets represent an exclusion criterium for septum expansion with osseodensification. This would either indicate the placement of ultra-wide implants or a delayed placement approach. In fact, our results showed that osseodensification allowed immediate implant placement in the first three categories (S-I, S-II, and S-III) of this new classification, with adequate implant stability.

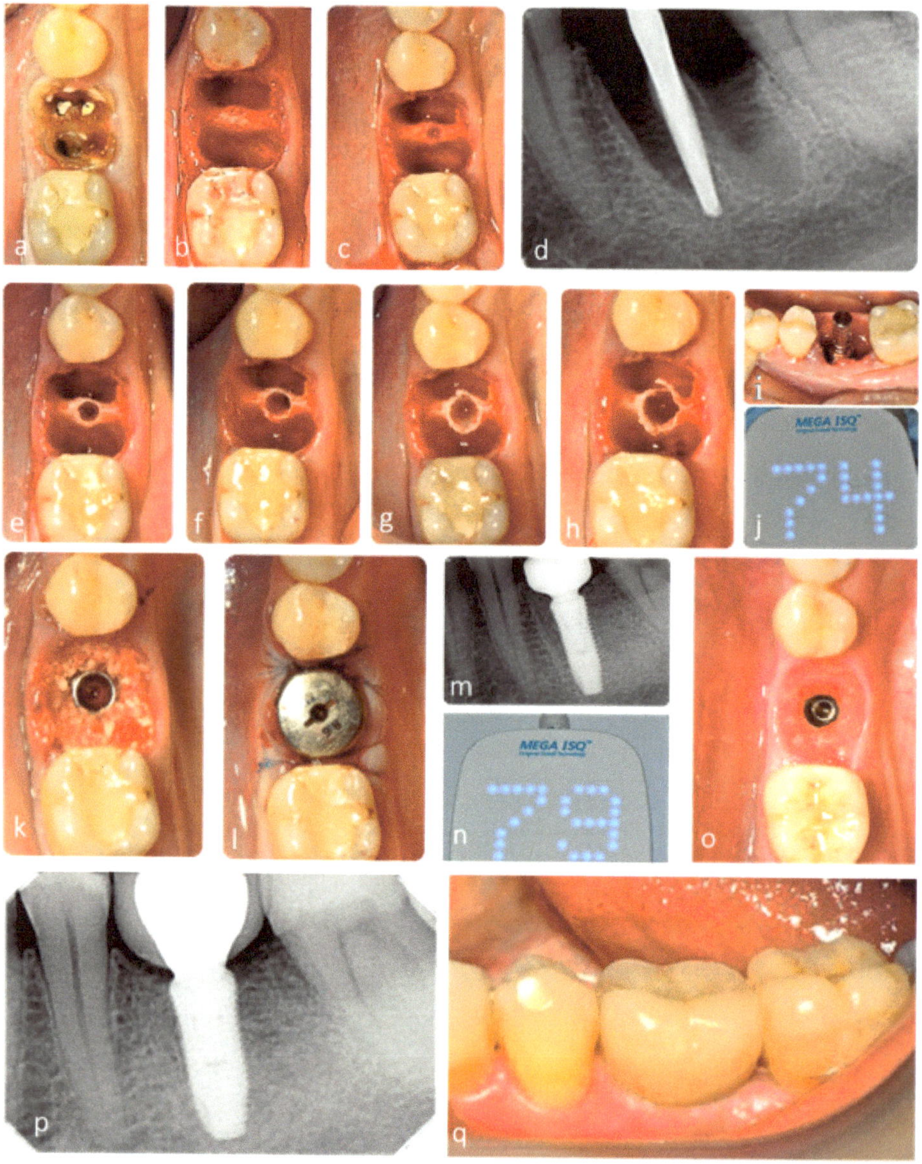

Figure 9. Representative clinical case with 4-year clinical and radiographic follow-up. (**a**) Initial radiograph with 4 mm of residual alveolar bone height. (**a**) Clinical situation at baseline. (**b**) Occlusal view after gentle tooth extraction with maintenance of interradicular septum. (**c–h**) Septum expansion after sequential instrumentation with osseodensification. (**d**) Radiograph of densifying bur VT1525 (2.0) in interradicular septum. (**i**,**j**) ISQ measurement after implant placement (primary stability). (**k–m**) Grafting of the gap and socket sealing with large healing abutment. (**n**) ISQ measurement after osseointegration period (secondary stability). (**o**) Contour maintenance after healing. (**p**,**q**) Clinical and radiographic follow-up after 4 years.

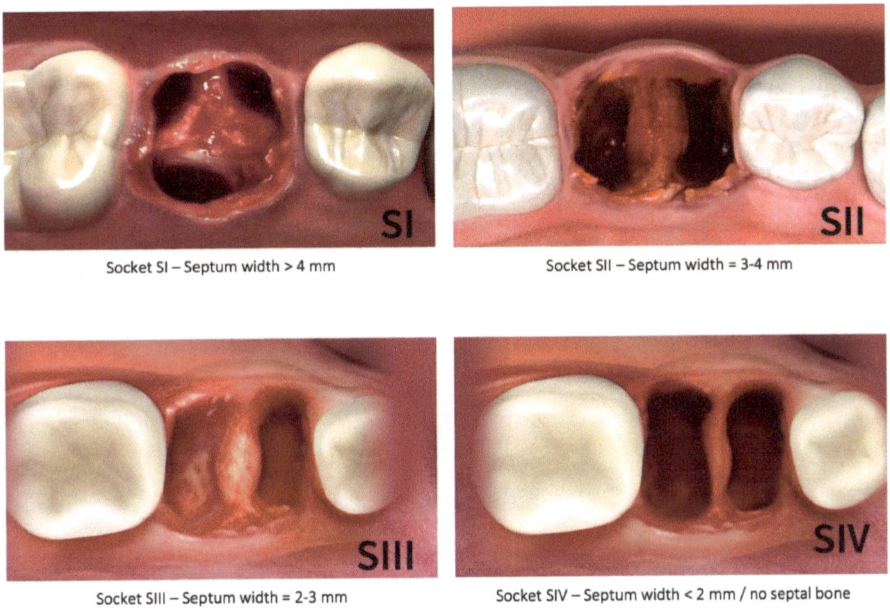

Figure 10. New molar socket classification according to the initial interradicular septum width. (SI—septum width > 4 mm; SII—septum width = 3–4 mm; SIII—septum width = 2–3 mm; SIV—septum width < 2 mm/no septal bone).

5. Conclusions

This up-to-5-year follow-up retrospective study showed that osseodensification is a viable and predictable method for interradicular septum expansion and immediate implant placement with adequate stability in molar extraction sockets. Furthermore, it allowed the introduction of a new molar socket classification, based on the available septum width prior to instrumentation. In the future, well-designed controlled clinical studies are needed to confirm these results and further explore the potential advantages of this site preparation technique.

Author Contributions: Conceptualization, S.B., J.G., S.H.; methodology, S.B., J.G., S.H.; software, J.G., S.H.; validation, J.G., J.J.M., R.N.; formal analysis, J.G., S.H.; investigation, S.B.; J.G., S.H., C.S., Z.M.; resources, S.H.; data curation, J.G., S.H.; writing—original draft preparation, J.G., S.H.; writing—review and editing, J.G., S.H., J.J.M., R.N.; visualization, J.G., S.H.; supervision, J.G., S.H., J.J.M., R.N. All authors have read and agreed to the published version of the manuscript.

Funding: This research received funding from the Foundation for Science and Technology, I.P., ID UIDB/04585/2020.

Institutional Review Board Statement: This study was declared exempt from the appropriate Institutional Review Board.

Informed Consent Statement: Informed consent was obtained from all subjects involved in the study.

Data Availability Statement: Data can be requested to the corresponding author upon reasonable request.

Conflicts of Interest: S.H. developed the novel osseodensification technique and invented the patented multifluted densifying burs that were utilized. None of the remaining authors report any conflict of interest.

References

1. Schulte, W.; Heimke, G. The Tübinger immediate implant. *Quintessenz* **1976**, *27*, 17–23.
2. Lazzara, R. Immediate implant placement into extraction sites: Surgical and restorative advantages. *Int. J. Periodontics Restor. Dent.* **1989**, *9*, 332–343.
3. Chen, S.; Wilson, T.J.; Hämmerle, C. Immediate or early placement of implants following tooth extraction: Review of biologic basis, clinical procedures, and outcomes. *Int. J. Oral Maxillofac. Implant.* **2004**, *19*, 12–25.
4. Wagenberg, B.; Froum, S. A retrospective study of 1925 consecutively placed immediate implants from 1988 to 2004. *Int. J. Oral Maxillofac. Implant.* **2006**, *21*, 70–80.
5. Schwartz-Arad, D.; Chaushu, G. The Ways and Wherefores of Immediate Placement of Implants into Fresh Extraction Sites: A Literature Review. *J. Periodontol.* **1997**, *68*, 915–923. [CrossRef] [PubMed]
6. Smith, R.B.; Tarnow, D.P.; Sarnachiaro, G. Immediate Placement of Dental Implants in Molar Extraction Sockets: An 11-Year Retrospective Analysis. *Compend. Contin. Educ. Dent.* **2019**, *40*, 166–170.
7. Ketabi, M.; Deporter, D.; Atenafu, E.G. A Systematic Review of Outcomes Following Immediate Molar Implant Placement Based on Recently Published Studies. *Clin. Implant. Dent. Relat. Res.* **2016**, *18*, 1084–1094. [CrossRef] [PubMed]
8. Ragucci, G.M.; Elnayef, B.; Criado-Cámara, E.; Del Amo, F.S.-L.; Hernández-Alfaro, F. Immediate implant placement in molar extraction sockets: A systematic review and meta-analysis. *Int. J. Implant. Dent.* **2020**, *6*, 40. [CrossRef]
9. Schwartz-Arad, D.; Grossman, Y.; Chaushu, G. The Clinical Effectiveness of Implants Placed Immediately into Fresh Extraction Sites of Molar Teeth. *J. Periodontol.* **2000**, *71*, 839–844. [CrossRef]
10. Fugazzotto, P.A. Implant Placement at the Time of Mandibular Molar Extraction: Description of Technique and Preliminary Results of 341 Cases. *J. Periodontol.* **2008**, *79*, 737–747. [CrossRef]
11. Walker, L.R.; Morris, G.A.; Novotny, P.J. Implant insertional torque values predict outcomes. *J. Oral Maxillofac. Surg.* **2011**, *69*, 1344–1349. [CrossRef]
12. Bavetta, G.; Bavetta, G.; Randazzo, V.; Cavataio, A.; Paderni, C.; Grassia, V.; Dipalma, G.; Gargiulo Isacco, C.; Scarano, A.; De Vito, D.; et al. A retrospective study on insertion torque and implant stability quotient (isq) as stability parameters for immediate loading of implants in fresh extraction sockets. *Biomed. Res. Int.* **2019**, *2019*, 9720419. [CrossRef] [PubMed]
13. Atieh, M.A.; Payne, A.G.T.; Duncan, W.J.; de Silva, R.K.; Cullinan, M.P. Immediate placement or immediate restoration/loading of single implants for molar tooth replacement: A systematic review and meta-analysis. *Int. J. Oral Maxillofac. Implant.* **2010**, *25*, 401–415.
14. Smith, R.B.; Tarnow, D.P. Classification of Molar Extraction Sites for Immediate Dental Implant Placement: Technical Note. *Int. J. Oral Maxillofac. Implant.* **2013**, *28*, 911–916. [CrossRef]
15. Marquezan, M.; Osório, A.; Sant'Anna, E.; Souza, M.M.; Maia, L. Does bone mineral density influence the primary stability of dental implants? A systematic review. *Clin. Oral Implant. Res.* **2012**, *23*, 767–774. [CrossRef]
16. Farronato, D.; Manfredini, M.; Stocchero, M.; Caccia, M.; Azzi, L.; Farronato, M. Influence of Bone Quality, Drilling Protocol, Implant Diameter/Length on Primary Stability: An In Vitro Comparative Study on Insertion Torque and Resonance Frequency Analysis. *J. Oral Implant.* **2020**, *46*, 182–189. [CrossRef]
17. Javed, F.; Ahmed, H.; Crespi, R.; Romanos, G. Role of primary stability for successful osseointegration of dental implants: Factors of influence and evaluation. *Interv. Med. Appl. Sci.* **2013**, *5*, 162–167. [CrossRef] [PubMed]
18. Greenstein, G.; Cavallaro, J. Implant Insertion Torque: Its Role in Achieving Primary Stability of Restorable Dental Implants. *Compend. Contin. Educ. Dent.* **2017**, *38*, 88–95.
19. Hattingh, A.; De Bruyn, H.; Vandeweghe, S. A retrospective study on ultra-wide diameter dental implants for immediate molar replacement. *Clin. Implant. Dent. Relat. Res.* **2019**, *21*, 879–887. [CrossRef]
20. Hattingh, A.; Hommez, G.; De Bruyn, H.; Huyghe, M.; Vandeweghe, S. A prospective study on ultra-wide diameter dental implants for immediate molar replacement. *Clin. Implant. Dent. Relat. Res.* **2018**, *20*, 1009–1015. [CrossRef] [PubMed]
21. Hattingh, A.; De Bruyn, H.; Ackermann, A.; Vandeweghe, S. Immediate Placement of Ultrawide-Diameter Implants in Molar Sockets: Description of a Recommended Technique. *Int. J. Periodontics Restor. Dent.* **2018**, *38*, 17–23. [CrossRef] [PubMed]
22. Huwais, S.; Meyer, E. A Novel Osseous Densification Approach in Implant Osteotomy Preparation to Increase Biomechanical Primary Stability, Bone Mineral Density, and Bone-to-Implant Contact. *Int. J. Oral Maxillofac. Implant.* **2017**, *32*, 27–36. [CrossRef] [PubMed]
23. Trisi, P.; Berardini, M.; Falco, A.; Podaliri Vulpiani, M. New osseodensification implant site preparation method to increase bone density in low-density bone: In vivo evaluation in sheep. *Implant. Dent.* **2016**, *25*, 24–31. [CrossRef]
24. Lahens, B.; Neiva, R.; Tovar, N.; Alifarag, A.M.; Jimbo, R.; Bonfante, E.A.; Bowers, M.M.; Cuppini, M.; Freitas, H.; Witek, L.; et al. Biomechanical and histologic basis of osseodensification drilling for endosteal implant placement in low density bone. An experimental study in sheep. *J. Mech. Behav. Biomed. Mater.* **2016**, *63*, 56–65. [CrossRef] [PubMed]
25. Alifarag, A.M.; Lopez, C.D.; Neiva, R.F.; Tovar, N.; Witek, L.; Coelho, P.G. Atemporal osseointegration: Early biomechanical stability through osseodensification. *J. Orthop. Res.* **2018**, *36*, 2516–2523. [CrossRef]
26. Kold, S.; Bechtold, J.E.; Ding, M.; Chareancholvanich, K.; Rahbek, O.; Søballe, K. Compacted cancellous bone has a spring-back effect. *Acta Orthop. Scand.* **2003**, *74*, 591–595. [CrossRef]
27. Lopez, C.D.; Alifarag, A.M.; Torroni, A.; Tovar, N.; Diaz-Siso, J.R.; Witek, L.; Rodriguez, E.D.; Coelho, P.G. Osseodensification for enhancement of spinal surgical hardware fixation. *J. Mech. Behav. Biomed. Mater.* **2017**, *69*, 275–281. [CrossRef]

28. Bergamo, E.T.; Zahoui, A.; Barrera, R.B.; Huwais, S.; Coelho, P.G.; Karateew, E.D.; Bonfante, E.A. Osseodensification effect on implants primary and secondary stability: Multicenter controlled clinical trial. *Clin. Implant. Dent. Relat. Res.* **2021**, *23*, 317–328. [CrossRef]
29. Koutouzis, T.; Huwais, S.; Hasan, F.; Trahan, W.; Waldrop, T.; Neiva, R. Alveolar Ridge Expansion by Osseodensification-Mediated Plastic Deformation and Compaction Autografting: A Multicenter Retrospective Study. *Implant. Dent.* **2019**, *28*, 349–355. [CrossRef]
30. Machado, R.; da Gama, C.; Batista, S.; Rizzo, D.; Valiense, H.; Moreira, R. Tomographic and clinical findings, pre-, trans-, and post-operative, of osseodensification in immediate loading. *Int. J. Growth Factors Stem Cells Dent.* **2018**, *1*, 101–105. [CrossRef]
31. Neiva, R.; Tanello, B.; Duarte, W.; Coelho, P.G.; Witek, L.; Silva, F. Effects of osseodensification on Astra TX and EV implant systems. *Clin. Oral Implant. Res.* **2018**, *29*, 444. [CrossRef]
32. Tanello, B.; Neiva, R.; Huwais, S. Osseodensification Protocols for Enhancement of Primary and Secondary Implant Stability- A Retrospective 5-year follow-up Multi-center Study. *Clin. Oral Implant. Res.* **2019**, *30*, 414. [CrossRef]
33. Gaspar, J.; Esteves, T.; Gaspar, R.; Rua, J.; João Mendes, J. Osseodensification for implant site preparation in the maxilla- a prospective study of 97 implants. *Clin. Oral Implants Res.* **2018**, *29*, 163. [CrossRef]
34. Kumar, B.; Narayan, V. Minimally invasive crestal approach sinus floor elevation using Densah burs, and Hydraulic lift utilising putty graft in cartridge delivery. *Clin. Oral Implants Res.* **2017**, *28*, 203.
35. Huwais, S.; Mazor, Z.; Ioannou, A.; Gluckman, H.; Neiva, R. A Multicenter Retrospective Clinical Study with Up-to-5-Year Follow-up Utilizing a Method that Enhances Bone Density and Allows for Transcrestal Sinus Augmentation Through Compaction Grafting. *Int. J. Oral Maxillofac. Implant.* **2018**, *33*, 1305–1311. [CrossRef] [PubMed]
36. Von Elm, E.; Altman, D.G.; Egger, M.; Pocock, S.J.; Gøtzsche, P.C.; Vandenbroucke, J.P. The Strengthening the Reporting of Observational Studies in Epidemiology (STROBE) statement: Guidelines for reporting observational studies. *Lancet* **2007**, *370*, 1453–1457. [CrossRef]
37. Tarnow, D.P.; Chu, S. Human histologic verification of osseointegration of an immediate implant placed into a fresh extraction socket with excessive gap distance without primary flap closure, graft, or membrane: A case report. *Int. J. Periodontics Restor. Dent.* **2011**, *31*, 515–521.
38. Tarnow, D.P.; Chu, S.J.; Salama, M.A.; Stappert, C.F.J.; Salama, H.; Garber, D.A.; Sarnachiaro, G.O.; Sarnachiaro, E.; Luis Gotta, S.; Saito, H. Flapless Postextraction Socket Implant Placement in the Esthetic Zone: Part 1. The Effect of Bone Grafting and/or Provisional Restoration on Facial-Palatal Ridge Dimensional Change—A Retrospective Cohort Study. *Int. J. Periodontics Restor. Dent.* **2014**, *34*, 323–331. [CrossRef]
39. Padhye, N.M.; Shirsekar, V.; Bhatavadekar, N.B. Three-Dimensional Alveolar Bone Assessment of Mandibular First Molars with Implications for Immediate Implant Placement. *Int. J. Periodontics Restor. Dent.* **2020**, *40*, e163–e167. [CrossRef]
40. Hayacibara, R.M.; Gonçalves, C.S.; Garcez-Filho, J.; Magro-Filho, O.; Esper, H.; Hayacibara, M. The success rate of immediate implant placement of mandibular molars: A clinical and radiographic retrospective evaluation between 2 and 8 years. *Clin. Oral Implant. Res.* **2013**, *24*, 806–811. [CrossRef]
41. Peñarrocha-Oltra, D.; Demarchi, C.; Maestre-Ferrín, L.; Peñarrocha-Diago, M.; Peñarrocha-Diago, M. Comparison of immediate and delayed implants in the maxillary molar region: A retrospective study of 123 implants. *Int. J. Oral Maxillofac. Implant.* **2012**, *27*, 604–610.
42. Smith, R.B.; Rawdin, S.B.; Kagan, V. Influence of Implant-Tooth Proximity on Incidence of Caries in Teeth Adjacent to Implants in Molar Sites: A Retrospective Radiographic Analysis of 300 Consecutive Implants. *Compend. Contin. Educ. Dent.* **2020**, *41*, e1–e5. [PubMed]

Journal of
Functional Biomaterials

MDPI

Article

Physical/Mechanical and Antibacterial Properties of Orthodontic Adhesives Containing Calcium Phosphate and Nisin

Supachai Chanachai [1], Wirinrat Chaichana [1], Kanlaya Insee [1], Sutiwa Benjakul [1], Visakha Aupaphong [2] and Piyaphong Panpisut [3,4,*]

1 Division of Orthodontics, Faculty of Dentistry, Thammasat University, Pathum Thani 12120, Thailand; supachai.chanachai@gmail.com (S.C.); fhunwirin@outlook.com (W.C.); ikanlaya@staff.tu.ac.th (K.I.); caredentist@hotmail.com (S.B.)
2 Division of Oral Biology, Faculty of Dentistry, Thammasat University, Pathum Thani 12120, Thailand; aupaphon@staff.tu.ac.th
3 Division of Restorative Dentistry, Thammasat University, Pathum Thani 12120, Thailand
4 Thammasat University Research Unit in Dental and Bone Substitute Biomaterials, Thammasat University, Pathum Thani 12120, Thailand
* Correspondence: panpisut@staff.tu.ac.th

Abstract: Enamel demineralization around orthodontic adhesive is a common esthetic concern during orthodontic treatment. The aim of this study was to prepare orthodontic adhesives containing monocalcium phosphate monohydrate (MCPM) and nisin to enable mineralizing and antibacterial actions. The physicomechanical properties and the inhibition of *S. mutans* growth of the adhesives with added MCPM (5, 10 wt %) and nisin (5, 10 wt %) were examined. Transbond XT (Trans) was used as the commercial comparison. The adhesive containing a low level of MCPM showed significantly higher monomer conversion (42–62%) than Trans (38%) ($p < 0.05$). Materials with additives showed lower monomer conversion ($p < 0.05$), biaxial flexural strength ($p < 0.05$), and shear bond strength to enamel than those of a control. Additives increased water sorption and solubility of the experimental materials. The addition of MCPM encouraged Ca and P ion release, and the precipitation of calcium phosphate at the bonding interface. The growth of *S. mutans* in all the groups was comparable ($p > 0.05$). In conclusion, experimental orthodontic adhesives with additives showed comparable conversion but lesser mechanical properties than the commercial material. The materials showed no antibacterial action, but exhibited ion release and calcium phosphate precipitation. These properties may promote remineralization of the demineralized enamel.

Keywords: orthodontic adhesives; monocalcium phosphate monohydrate; nisin; monomer conversion; water sorption; water solubility; biaxial flexural strength; shear bond strength; ion release; calcium phosphate precipitation; antibacterial; *Streptococcus mutans*

1. Introduction

The most common complication during fixed orthodontic treatment is white spot lesions around the bracket base. The prevalence is approximately 25–30%, which can vary depending on detection criteria across studies [1]. The lesions are associated with loss of balance between mineral loss (demineralization) and gain (remineralization) from the acid produced by a dysbiotic biofilm around the fixed appliances [2]. This results in the net demineralization of enamel and subsurface porosities, which appears as whitish lesions. If the lesions are left untreated, carious lesions may progress and become uncleanable cavities causing severe infection and pain that require the intervention of orthodontic treatment. The most commonly used orthodontic adhesive is resin-based composite due to its excellent optical properties and strong adhesion to enamel. The peripheral area of the adhesive excesses promotes plaque retention around the appliances [3,4]. The main limitation of the current resin composite orthodontic adhesives is the lack of remineralizing and antibacterial properties. This may lead to the continuation of tooth demineralization and the progression of carious lesions.

Various ion-releasing fillers such as fluoride compounds [5], bioactive glass [6], or nanoparticles of calcium phosphate [7] have been added into the resin composite-based orthodontic adhesives to enhance the remineralizing action of the materials. However, strong clinical evidence to support their benefits is limited. Calcium and phosphate ions are essential to enable a suitable saturated condition for the precipitation of hydroxyapatite ($Ca_{10}(PO_4)_6OH_2$), which is the main inorganic structure in tooth minerals [8]. Materials with the ability to promote apatite formation are expected to enable bottom-up or top-down remineralization in the demineralized tooth structure. Previous studies have incorporated monocalcium phosphate monohydrate (MCPM; 10–40 wt %) into dental composites to promote mineralizing actions for the materials [9–13]. MCPM is a commercially available calcium phosphate compound with excellent ion-releasing and hydroxyapatite formation abilities [14]. Previous studies showed that an increase in MCPM (from 10 to 40 wt %) reduced the flexural strength of the composite [15]. Hence, a low level of MCPM may be required to minimize the negative effects on the physical or mechanical properties of the materials.

Various antibacterial fillers were incorporated into the adhesives, such as chlorhexidine [16] to promote the antibacterial actions of the materials. The concerns with chlorhexidine are the risk of causing a severe allergic reaction [17,18] and the development of antibiotic resistance [19,20], which is a current global health threat. Nisin is the antimicrobial cationic peptide produced from the *Lactococcus* and *Streptococcus* species. It has been approved for use as a biological preservative due to its nontoxicity and its antimicrobial actions. Nisin also demonstrated strong broad-spectrum antimicrobial activity, low tendency to be resistant to bacteria, and low toxicity to human cells at the bactericidal concentration [21]. The proposed antibacterial actions of nisin are mainly from the interaction between anionic phospholipids in the bacterial cell membrane, leading to bacterial cell lysis [22,23]. Previous studies demonstrated that the addition of nisin into experimental dentin bonding agents inhibited the growth of both *S. mutans* monospecific biofilm and saliva-derived multispecies biofilm [24,25]. The addition of nisin also showed no detrimental effects on the bond strength and degree of monomer conversion of the materials. The use of nisin in orthodontic adhesives has not yet been investigated.

The aim of the current study was, therefore, to prepare new orthodontic adhesives containing MCPM (at a low concentration of 5 or 10 wt %) and nisin, and to test their physicochemical properties. We tested the effect of rising MCPM and nisin concentrations on the degree of monomer conversion, biaxial flexural strength and modulus, water sorption and solubility, shear bond strength to enamel, calcium phosphate precipitation, and the growth of *S. mutans*. The first hypothesis was that the additives should not exhibit significant effects on the tested properties of the materials. The second hypothesis was that the properties of the experimental adhesives tested in the current study should not be significantly different to those of the commercial material.

2. Materials and Methods

2.1. Materials and Methods

The liquid phase contained 70 wt % urethane dimethacrylate (UDMA, Sigma-Aldrich, St. Louis, MO, USA), 25 wt % triethylene glycol dimethacrylate (TEGMDA, Sigma-Aldrich, St. Louis, MO, USA), 4 wt % 2-hydroxyethyl methacrylate (HEMA, Sigma-Aldrich, St. Louis, MO, USA), and 1 wt % camphorquinone (CQ, Sigma-Aldrich, St. Louis, MO, USA). The powder phase contained silanated boroaluminosilicate glass (particle diameter of 0.7 and 7 µm, Esstech, Essington, PA, USA), monocalcium phosphate monohydrate (MCPM, a particle diameter of 10 µm, Old Bethpage, NY, USA), and nisin (Nisin Z, Handary, Evere, Belgium). The formulations of experimental orthodontic adhesives are presented in Table 1.

Table 1. Composition of powder phase (wt %) of experimental orthodontic adhesives.

Formulations	Boroaluminosilicate Glass (7 μm)	Boroaluminosilicate Glass (0.7 μm)	MCPM	Nisin
M10N10	40	40	10	10
M10N5	42.5	42.5	10	5
M5N10	42.5	42.5	5	10
M5N5	45	45	5	5
M0N0	50	50	0	0

The composite paste was prepared by mixing powder and liquid phase using a powder to liquid ratio of 3:1. The mixed composites were placed into a composite syringe (MIXPAC 1 mL syringe, Sulzer Mixpac AG, Haag, Switzerland). The commercial orthodontic adhesive (Transbond XT, 3M ESPE, St. Paul, MN, USA) was used as the commercial control (Table 2).

Table 2. Composition of commercial orthodontic adhesive (Transbond XT, 3M ESPE, St. Paul, MN, USA). Actual composition is protected as a trade secret of the manufacturer.

Composition	Amount (wt %)
Silane-treated quartz	70–80
Bisphenol A diglycidyl ether dimethacrylate (Bis-GMA)	10–20
Bisphenol A bis (2-hydroxyethyl ether) dimethacrylate	5–10
Silane treated silica	<2
Diphenyliodonium hexafluorophosphate	<1

2.2. Degree of Monomer Conversion

Composites (n = 5) were placed in the metal circlip (1 mm in thickness and 10 mm in diameter, Springmaster Ltd., Redditch, UK) on the diamond of a Fourier-transform infrared spectroscope (FTIR, Nicolet iS5, Thermo Fisher Scientific, Waltham, MA, USA) equipped with attenuated total reflection (ATR, iD7 ATR, Thermo Fisher Scientific, Waltham, MA, USA). Composites were covered with an acetate sheet. Then, specimens were light-cured for 20 s using an LED light-curing unit (irradiance of 1200 mW/cm^2, SmartLite Focus Pen Style, DENTSPLY Sirona, York, PA, USA). FTIR spectra in the region of 700–4000 cm^{-1} from the bottom of the specimens before and after curing were recorded. The degree of monomer conversion (DC, %) was then obtained using the following equation [12].

$$D_c = \frac{100(\Delta A_0 - \Delta A_t)}{\Delta A_0} \quad (1)$$

where ΔA_0 and ΔA_t are the absorbance of the C–O peak (1320 cm^{-1}) [26] above the baseline at 1335 cm^{-1} before and after curing at time t, respectively.

2.3. Biaxial Flexural Strength (BFS) and Biaxial Flexural Modulus (BFM)

Disc specimens were prepared (n = 8). Composites were placed into a metal circlip (10 mm in diameter and 1 mm in thickness), and covered with an acetate sheet and glass slides on top and bottom surfaces. They were cured by an LED light-curing unit for 20 s on both sides using circular motions. Specimens were left for 24 h at 25 ± 1 °C to allow for the completion of polymerization. Then, they were immersed in 10 mL of deionized water and incubated at 37 °C for 24 h. Biaxial flexural strength (BFS) testing was conducted using a ball-on-ring testing jig under a mechanical testing frame (AGSX, Shimadzu, Kyoto, Japan). The test was performed using a 500 N load cell with a crosshead speed of 1 mm/min. The force was applied until the specimen failed. BFS (Pa) was calculated using the following equation [11].

$$BFS = \frac{F}{d^2}\left\{(1+a)\left[0.485\ln\left(\frac{r}{d}\right)+0.52\right]+0.48\right\} \quad (2)$$

where F is the failure load (N), d is the thickness of the sample (m), r is the radius of circular support (mm), and a is Poison's ratio (0.3). Additionally, the biaxial flexural modulus (BFM, Pa) was calculated using the following equation:

$$BFM = \left(\frac{\Delta H}{\Delta W_c}\right) \times \left(\frac{\beta_c d^2}{q^3}\right) \quad (3)$$

where $\frac{\Delta H}{\Delta W_c}$ is the rate of change of the load with regard to the central deflection or gradient of force versus the displacement curve (N/m), β_c is the center deflection junction (0.5024), and q is the ratio of the support radius to the radius of the disc.

2.4. Water Sorption and Solubility

Assessment of the water sorption and solubility of the materials was performed according to BS EN ISO 4049:2019, Dentistry—polymer-based restorative materials [27]. Disc specimens were prepared (n = 6) and placed in a desiccator at 37 ± 1 °C for 22 h. Specimens were transferred to a desiccator and placed in an incubator with controlled temperature at 37 ± 1 °C for 22 h. Then, specimens were removed from the first desiccator and transferred to the second desiccator (25 ± 1 °C) for 2 h. The weight of specimens was measured using a four-figure balance. These steps were repeated until a constant mass or m_1 was obtained [27].

Specimens were then placed in 10 mL of deionized water at 37 ± 1 °C for up to 4 weeks. The specimen mass was then recorded until a constant mass (m_2) was obtained. Specimens were then reconditioned following the steps described above for m_1. Reconditioning was performed until a constant mass (m_3) was obtained. Water sorption (W_{SP}, g/m^3) and water solubility (W_{SL}, g/m^3) were calculated using the following equations.

$$W_{SP} = \frac{m_2 - m_3}{V} \quad (4)$$

$$W_{SL} = \frac{m_1 - m_3}{V} \quad (5)$$

where m_1 is the conditioned mass of the specimen (g), m_2 is the mass of the specimen after immersion in water for 4 weeks (g), m_3 is the reconditioned mass of the specimen after immersion in water (g), and v is the volume of the specimen (m^3).

2.5. Enamel Shear Bond Strength

Collecting extracted teeth was approved by the Ethics Review Subcommittee Board for Human Research Involving Sciences, Thammasat University, No. 3 (Faculty of Health Sciences and Science and Technology, date of issue: 11 November 2020). The thirty extracted premolars with no visible caries or noncarious lesions were collected at Thammasat University Hospital, Pathum Thani, Thailand. Teeth were kept in 0.1% thymol solution at room temperature for less than 30 days prior to the test.

Teeth (n = 5) were cleaned and assessed for defects under a stereomicroscope. The root was cut at 2 mm under the cervical line. The buccal surface was then cleaned with pumice and water for 15 s. The surface was etched with 37% phosphoric acid (Transbond^{TM} XT etching gel; 3M Unitek, Monrovia, CA, USA) for 15 s, followed by rinsing with water for 15 s and air-drying with a three-way syringe. The etched surface was applied with a primer (Transbond^{TM} XT Light Cure Orthodontic Primer; 3M Unitek, Monrovia, CA, USA) for 10 s and air-dried. Experimental and commercial adhesives were then placed onto the tooth surface. Premolar brackets (GEMINI MBT 0.022 Twin, 3M Unitek, Monrovia, CA, USA) were placed on the adhesive. Excess adhesive was removed. Then, the specimen was light-cured using an LED light-curing unit for 10 s on each side (mesial and distal) of the bracket. Specimens were embedded in a self-cure acrylic resin in a PVC tube (Figure 1A).

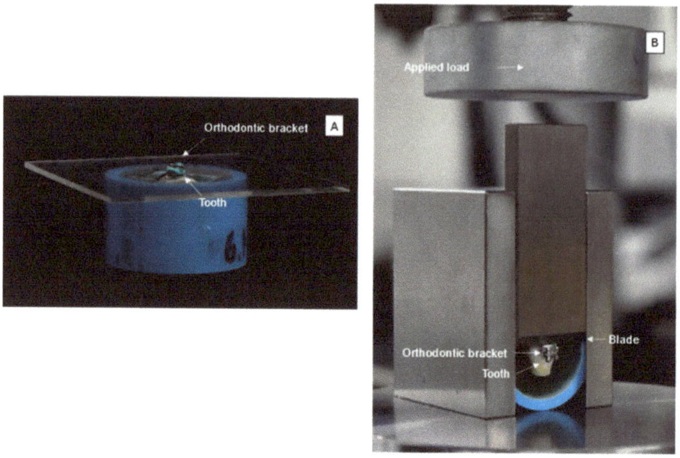

Figure 1. (**A**) Orthodontic bracket attached parallel to the buccal surface of a tooth. (**B**) Blade of SBS testing jig positioned at the interface between bracket and tooth surface.

Specimens from each group were immersed in artificial saliva for 24 h. Then, specimens were subjected to thermocycling between 5 and 55 °C for 500 cycles according to PD ISO/TS 11405:2015 (Dentistry—Testing of adhesion to tooth structure) [28]. Immersion time in each bath and dwell time were 30 and 10 s, respectively. Then, specimens were placed in a shear bond strength testing jig (Figure 1B). The knife-edge chisel was positioned at the interface between tooth and bracket. The jig was placed under the mechanical testing frame (AGSX, Shimadzu, Kyoto, Japan). Shear bond strength (SBS) testing was conducted using a 500 N load cell and a crosshead speed of 1 mm/min. Maximal load (F, Newton) before the debonding of the bracket was recorded. SBS (Pa) was then calculated using the following equation [29].

$$\mathrm{SBS} = \frac{\mathrm{F}}{\mathrm{A}} \quad (6)$$

where A is the area of the bonding surface of the bracket (m^2). Then, the adhesive remnant index was analyzed by examining the residual adhesive on the bracket under a stereomicroscope (10x magnification). The classification of the ARI index was as follows [30,31].

(1) Score 0: no adhesive remained on the enamel.
(2) Score 1: less than 50% of the adhesive remained on the enamel surface.
(3) Score 2: more than 50% of the adhesive remained on the enamel surface.
(4) Score 3: all adhesive remained on the enamel surface.

2.6. Calcium Phosphate Precipitation

The specimen was prepared according to Section 2.5 ($n = 1$). The specimens were immersed in 10 mL of artificial saliva and incubated at 37 °C for 24 h. Then, the bracket was debonded. The bonding interface of the detached bracket was sputter-coated with Au using a sputter-coating machine (Quoram Q150R ES®, East Sussex, UK) with a current of 23 mA for 45 s. The surface was then examined under a scanning electron microscope (SEM, JSM, 7800F, JEOL Ltd., Tokyo, Japan) to investigate the calcium phosphate precipitation. Additionally, energy dispersive X-ray analysis (EDX, -sight 6650 detector, Oxford Instruments, Abingdon, UK) was employed to analyze the elemental composition of the precipitation using a magnification of 1000–5000× and a beam voltage set at 10 kV [12].

2.7. Ion Release

The storage solution from the water sorption and solubility test (n = 3) at 4 weeks was collected for assessing the concentration of Ca and P ions. The collected solution was mixed with 3 vol % nitric acid. The standard calibration was performed using the instrument calibration standards. The ion concentration of Ca and P ions in the mixed solution was analyzed using inductively coupled plasma atomic emission spectroscopy (ICP-OES, Optima 8300, PerkinElmer, Waltham, MA, USA) [32]. The result was analyzed using Syngistix TM for ICL software version 2.0 (PerkinElmer, Waltham, MA, USA).

2.8. Fluence to S. mutans Growth

Streptococcus mutans (ATCC 25175) was inoculated in Mueller Hinton (MH) broth (BD Difco™ Mueller Hinton Broth, Thermo Fisher Scientific Inc., Göteborg, Sweden) using a 1:2 volume ratio of inoculum to broth. Tubes were incubated for 24 h at 37 °C in air enriched with 5% CO_2. The suspension of *S. mutans* was then adjusted by spectrophotometry at optical density (OD) of 600 nm. The concentration of bacterial suspension was diluted until the bacterial concentration of 2.5×10^5 cell/mL had been obtained.

Disc specimens were prepared (n = 3) and sterilized under UV irradiation for 30 min on the bottom and top surfaces [33]. Then, they were immersed in the tube containing the mixture between 2 mL of Mueller Hinton Broth and 1 mL of the suspension of *S. mutans*. The tube without a disc specimen was used as the control. Tubes were incubated at the controlled temperature of 37 °C in air enriched with 5% CO_2 for 48 h. Then, discs were removed. The suspension was vortexed for 30 s, followed by serial dilution until a bacterial concentration of 1×10^{-6} CFU/mL had been obtained. The suspension (200 µL) was then plated on the Mitis Salivarius agar. Plates were then incubated for 48 h at 37 °C under 5% CO_2 atmosphere. Colony-forming units (log CFU/mL) [34] were then counted under microscope and image analysis (ImageJ, National Institutes of Health, Bethesda, MD, USA).

2.9. Statistical Analysis

Numerical results reported in the study are mean and SD. Data were analyzed using Prism version 9.2 for macOS (GraphPad Software, San Diego, CA, USA). Data normality was analyzed using the Shapiro–Wilk test. For normally distributed results, one-way ANOVA followed by Tukey's post hoc multiple-comparison test were performed. Additionally, the Kruskal–Wallis test followed by multiple comparisons using the Dunn test was employed for non-normally distributed results. A chi-squared test was used to evaluate the ARI scores among the adhesive subgroups. Statistical significance was set at $p = 0.05$. The sample size used in each test was calculated using G*Power 3.1 software (University of Dusseldorf, Dusseldorf, Germany) using the results in published studies [9,12,35] and a pilot study. The result indicated that the sample size in each test gave power >0.95 at alpha = 0.05. Additionally, the effects of increasing MCPM and nisin concentrations on the tested properties were assessed using factorial analysis [11].

3. Results

3.1. Degree of Monomer Conversion

The highest and lowest monomer conversions were obtained from M0N0 (62.2 ± 0.4%) and Trans (37.9 ± 1.0%), respectively (Figure 2). The conversion of M0N0 was significantly higher than that of M10N10 (38.6 ± 0.4%), M10N5 (39.3 ± 1.4%), M5N10 (41.7 ± 0.7%), and M5N5 (41.6 ± 0.8%) ($p < 0.05$). The conversion of M5N5 and M5N10 was significantly higher than that of Trans, M10N10, and M10N5 ($p < 0.05$). Factorial analysis indicated that the increase in MCPM level from 5 to 10 wt % reduced the degree of monomer conversion by 4 ± 2%, while the effect from rising nisin was negligible.

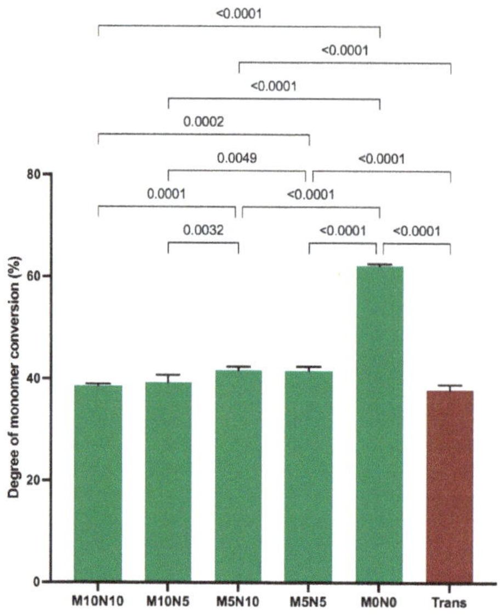

Figure 2. Degree of monomer conversion of all materials after light curing for 40 s. Error bars are SD (n = 5). Lines indicate p value.

3.2. *Biaxial Flexural Strength (BFS) and Modulus (BFM)*

The highest and lowest BFS were detected with M0N0 (220.0 ± 16.7 MPa) and M10N10 (109.3 ± 7.4 MPa), respectively (Figure 3A). The BFS of M0N0 was comparable to that of Trans (202.1 ± 19.2 MPa) (p = 0.0607). The BFS of Trans and M0N0 was significantly higher than that of M10N10, M10N5 (136.8 ± 11.5 MPa), M5N10 (109.3 ± 5.3 MPa), and M5N5 (145.0 ± 6.8 MPa) ($p < 0.05$).

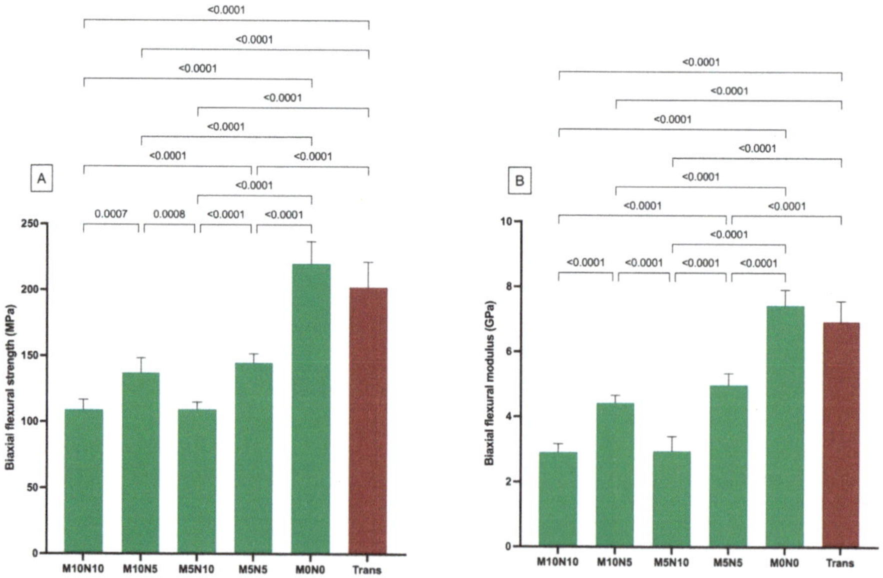

Figure 3. (**A**) Biaxial flexural strength and (**B**) biaxial flexural modulus of all materials after immersion in deionized water for 24 h. Error bars are SD (n = 8). Lines indicate $p < 0.05$.

For BFM (Figure 3B), M0N0 exhibited the highest BFM (7.4 ± 0.5 GPa). The BFM of M0N0 was also similar to that of Trans (6.9 ± 0.6 GPa) (p = 0.0607). The BFM of both Trans and M0N0 was significantly higher than that of M10N10 (2.9 ± 0.3 GPa), M10N5 (4.4 ± 0.2 GPa), M5N10 (2.9 ± 0.5 GPa), and M5N5 (5.0 ± 0.4 MPa) ($p < 0.05$).

Factorial analysis showed that the increase in nisin level reduced BFS and BFM by 22 ± 4% and 37 ± 9%, respectively. The effect of increasing the MCPM level was minimal.

3.3. Water Sorption (W_{SP}) and Water Solubility (W_{SL})

The highest and lowest W_{SP} were detected with M10N10 (234 ± 4 μg/mm^3) and Trans (12 ± 2 μg/mm^3), respectively (Figure 4A). The W_{SP} of M10N10 was significantly higher than that of M10N5 (178 ± 1 μg/mm^3), M5N10 (193 ± 2 μg/mm^3), M5N5 (134 ± 3 μg/mm^3), M0N0 (22 ± 2 μg/mm^3), and Trans ($p < 0.01$). Additionally, the W_{SP} of all groups were significantly different from each other ($p < 0.01$). Factorial analysis indicated that the increase in MCPM and nisin enhanced W_{SP} by 17 ± 14% and 23 ± 19%, respectively.

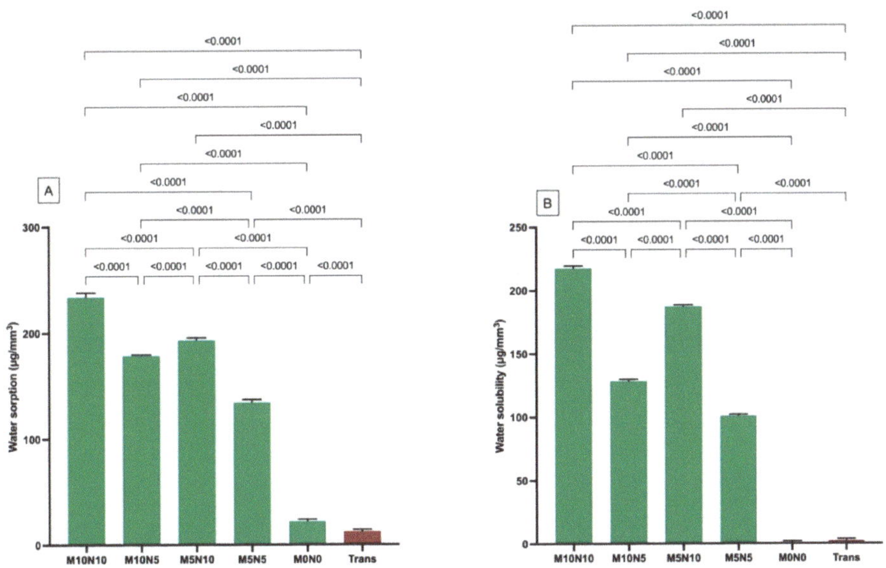

Figure 4. (**A**) Water sorption and (**B**) water solubility of materials upon immersion in deionized water for 4 weeks. Error bars are SD (n = 6). Lines indicate $p < 0.05$.

The highest and lowest W_{SL} were detected with M10N10 (217.6 ± 1.9 μg/mm^3) and M0N0 (0.7 ± 1.1 μg/mm^3), respectively (Figure 4B). M10N10 showed significantly higher W_{SL} than that of N10N5 (128.4 ± 1.4 μg/mm^3), M5N10 (187.5 ± 1.1 μg/mm^3), M5N5 (100.9 ± 1.0 μg/mm^3), M0N0, and Trans (2.0 ± 1.5 μg/mm^3) ($p < 0.01$). Factorial analysis indicated that the increase in MCPM and nisin enhanced W_{SP} by 13 ± 11% and 48 ± 40%, respectively.

3.4. Enamel Shear Bond Strength (SBS) and Adhesive Remnant Index (ARI) Score

The highest and lowest SBS were obtained from M0N0 (32 ± 3 MPa) and M5N10 (14 ± 6 MPa) (Figure 5). The SBS of M0N0 was significantly higher than that of M10N10 (19 ± 8 MPa) (p = 0.0455). Additionally, the SBS of M5N10 was significantly lower than that of M5N5 (29 ± 3 MPa), M0N0, and Trans (31 ± 8 MPa) ($p < 0.01$). The increase in nisin level reduced SBS by 19 ± 15%, while the increase in MCPM showed a negligible effect.

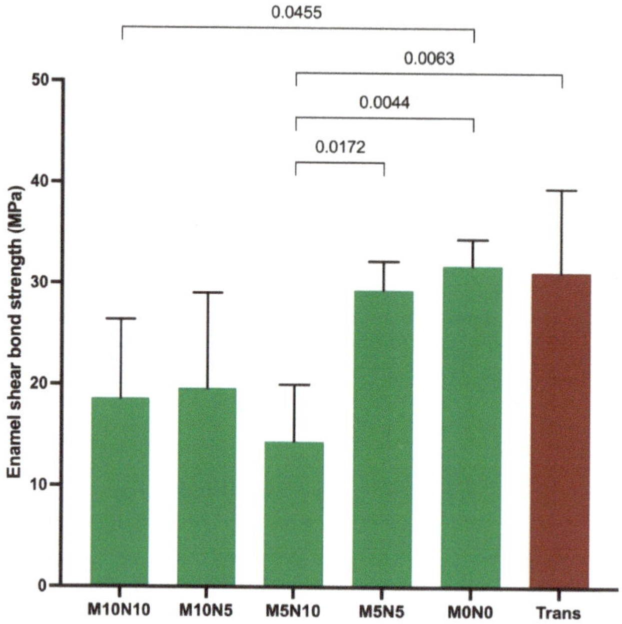

Figure 5. Enamel shear bond strength of all materials. Error bars are SD (n = 5). Lines indicate $p < 0.05$.

The distribution of the ARI score (Figure 6) among each group was significantly different ($p < 0.05$). The most common ARI scores observed for the experimental materials were 1 and 2. A score of 3 on the ARI index was only detected for Trans.

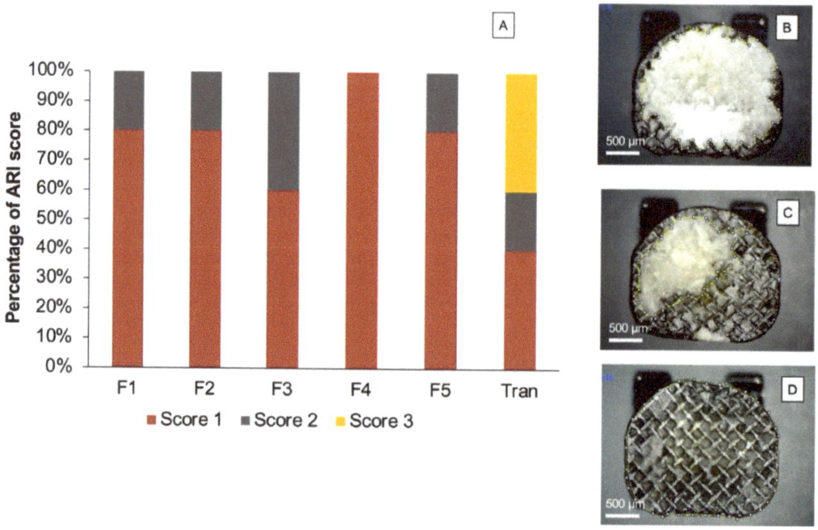

Figure 6. (**A**) Percentage of ARI score observed after SBS testing. Example of ARI score observed from randomly selected materials; (**B**) score 1, (**C**) score 2, and (**D**) score 3.

3.5. Calcium Phosphate Precipitation

The surface of the debonded brackets from the randomly selected specimens of M10N10, M10N5, M5N10, and M5N5 showed rod-shaped precipitation on the adhesives (Figure 7). EDX showed that the observed precipitate contained Ca and P (Figure 8). No precipitation was detected on the adhesive of M0N0 and Trans.

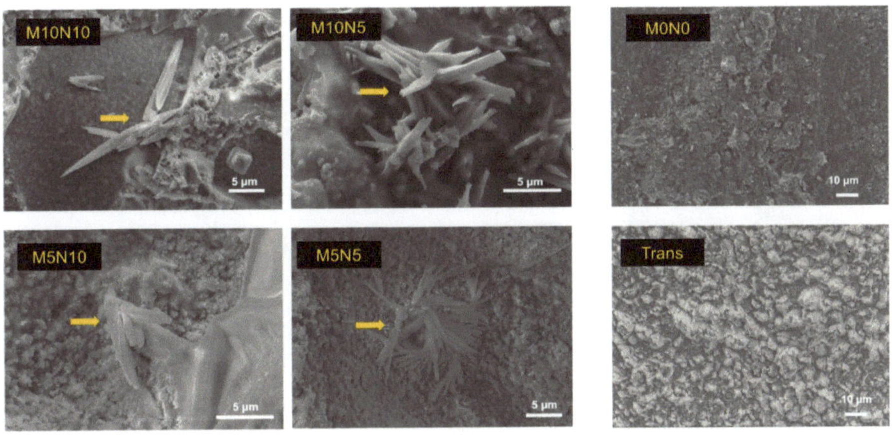

Figure 7. Surface of adhesives of representative specimen after debonding from enamel. Precipitates (arrows) were observed on M10N10, M10N5, M5N10, and M5N5. Precipitation was not detected on M0N0 and Trans.

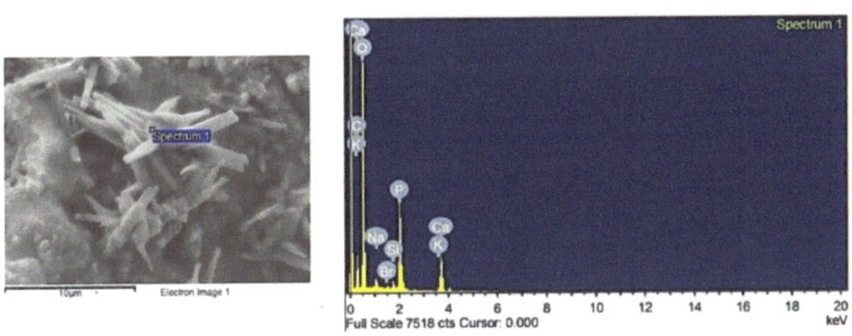

Element	Atomic %
C	21.93
O	63.50
Na	0.77
Si	0.48
P	7.49
K	0.16
Ca	5.53
Br	0.15

Figure 8. Example of EDX result obtained from precipitate detected on a representative specimen of M10N5.

3.6. Ion Release

The levels of Ca and P in M0N0 and Trans were under the detectable levels (<0.13 ppm). The highest and lowest Ca ion concentrations were detected in M10N10 (39.3 ± 0.6 ppm) and M5N5 (10.2 ± 0.8 ppm), respectively (Figure 9). The Ca ion concentration in M10N10 was significantly higher than that in M10N5 (30.8 ± 0.3 ppm), M5N10 (13.4 ± 0.4 ppm), and M5N5 ($p < 0.05$). Similarly, the highest and lowest P ion concentrations were obtained from M10N10 (80.3 ± 0.7 ppm) and M5N5 (20.2 ± 0.9 ppm), respectively. M10N10 showed significantly higher P ion concentration than M10N5 (62.8 ± 0.9 ppm), M5N10 (27.9 ± 0.3 ppm), and M5N5 ($p < 0.05$).

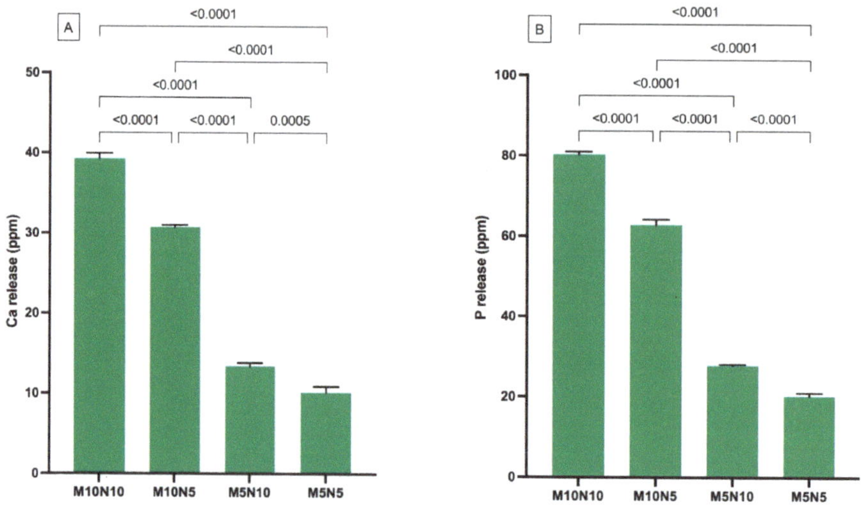

Figure 9. (**A**) Calcium and (**B**) phosphorus in storage solution at 4 weeks. Error bars are SD (n = 3). Lines indicate p values.

Additionally, factorial analysis showed that the increase in MCPM from 5 to 10 wt % increased Ca and P ion release by 199 ± 12% and 200 ± 9%, respectively. Furthermore, the increase in nisin also increased Ca and P ion release by 30 ± 6% and 33 ± 3%, respectively.

3.7. Influence on *S. mutans* Growth

The lowest and highest Log CFU/mL were detected with F5 (3.47 ± 0.42 Log CFU/mL) and F1 (3.21 ± 0.24 Log CFU/mL), respectively (Figure 10). Bacterial colonies of each group were numbered as follows. F2 (3.49 ± 0.27 Log CFU/mL), F3 (3.46 ± 0.24 Log CFU/mL), F4 (3.22 ± 0.24 Log CFU/mL), and Trans (3.35 ± 0.24 Log CFU/mL). However, no statistically significant difference was detected between groups ($p > 0.05$). The increase in MCPM and nisin also showed a negligible effect on the growth of *S. mutans*.

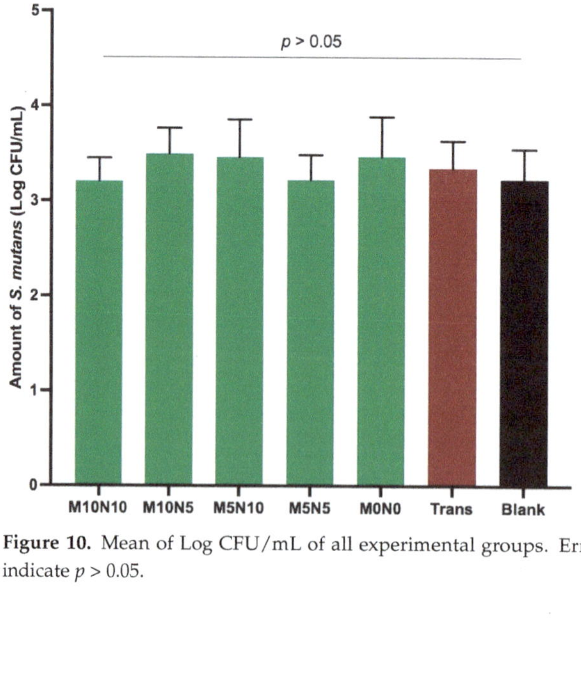

Figure 10. Mean of Log CFU/mL of all experimental groups. Error bars are SD (n = 3). Lines indicate $p > 0.05$.

4. Discussion

Experimental orthodontic adhesives containing MCPM and nisin were prepared. The effects of increasing MCPM and nisin from 5 to 10 wt % on the physical or mechanical properties and influence on the growth of *S. mutans* were assessed. The increase in the levels of the additives affected the degree of monomer conversion, water sorption and solubility, biaxial flexural strength and modulus, shear bond strength, and the ion release of the materials. Hence, the first hypothesis was rejected. The second hypothesis was also rejected, as the monomer conversion, biaxial flexural strength and modulus, water sorption and solubility, and ion release obtained from the experimental materials were not comparable to those of the commercial materials. The current study is an in vitro study. Hence, further in vivo or in situ experiments should be conducted to confirm the beneficial effects of the experimental orthodontic adhesives.

4.1. Degree of Monomer Conversion

A high degree of monomer conversion of orthodontic adhesives may help in reducing the risk of uncured monomer elution. Released monomers could be detected even after long-term immersion for up to 52 weeks [36]. Eluted monomers from orthodontic adhesives are slightly toxic to human gingival fibroblasts [37]. Additionally, the detection of BPA from commercial adhesives was reported, which could be due to impurities during Bis-GMA synthesis. Hence, UDMA was used as the base monomer in the experimental adhesives in the current study to avoid BPA-induced estrogenic activities [38]. Furthermore, Bis-GMA may promote sugar transport and the accumulation of intracellular polysaccharides in *S. mutans*, which may increase the cariogenicity of dental biofilm [39]. However, a study showed that UDMA also enhanced the tolerance of oxidative stress to *S. mutans* and favored biofilm development at the early stage [40].

The monomer conversion of Trans was comparable to that reported in the published study (~43%) [41]. The DC of the experimental adhesives was comparable to or higher than that of Trans. However, the minimal requirement for the DC of orthodontic adhesives has not yet been established. The degree of monomer conversion of the experimental materials without additive was within the range of that reported for restorative resin composites (~50–70%) [42]. The higher DC of M0N0 compared with Trans could be due to the differences in base monomers used in the materials. The use of low glass transition (T_g) temperature monomers could enhance the degree of monomer conversion for the polymer [43]. The T_g of UDMA (−35 °C) was lower than that of Bis-GMA (−8 °C) [44]. The additives significantly reduced the degree of monomer conversion of the materials by almost 20%. The addition of MCPM and nisin may increase in refractive index mismatch within the composites. This could potentially reduce light penetration into the bottom of the specimens. Additives also generally reduce the light intensity inside the specimen. This may have been because additives increased the refractive index mismatch in the adhesive [12,45,46]. This may consequently reduce light penetration in the material to the bottom surface of the adhesive.

4.2. Biaxial Flexural Strength and Modulus

The strength of orthodontic adhesive should be sufficiently high to ensure that the material can withstand the applied forces without bracket debonding. There is no standard requirement for resin composite orthodontic adhesives. The most relevant standard is BS ISO 4049: dentistry—polymer-based restorative materials for luting materials. Resin-based luting materials that require external energy for polymerization should exhibit flexural strength higher than 50 Mpa [27]. Experimental adhesives in all formations showed higher flexural strength than that required by the standard (109–220 Mpa), which may indicate that the materials would pass the requirement. However, a limitation of the current study is that the flexural strength testing was biaxial flexural strength testing, while the testing required by BS ISO 4049 is a three-point bending test. Advantages of the biaxial flexural test compared with the three-point bending test include smaller specimens, low technique sensitivity for specimen preparation, and low risk of undesirable failure at the edges of specimens due to flaws (edge failures) [47]. The 2017 guideline for material testing from the Academy of Dental Materials also indicated that the result from the biaxial flexural strength test was correlated with the results from the three-point bending test but with lower variation [48,49].

The mechanical properties of materials can be influenced by various factors, such as the composition of monomers, filler loading, and the type of filler [50,51]. The replacement of silane-treated glass fillers by hydrophilic and nonsilanized fillers led to a reduction in mechanical properties, which was in accordance with previous studies [15,52–54]. The lack of silanation may, however, reduce filler–matrix interaction, which could encourage crack propagation during strength testing [53,55]. Additives were not silanized to facilitate the reaction with water or release from the adhesives.

MCPM is highly soluble in water (solubility of 18 g/L), which can readily react with water to release essential ions for promoting mineral precipitation [56]. A previous study incorporated the high level of MCPM (10–20 wt %) to enhance the remineralizing effects of resin composites [11,15]. However, the use of such an MCPM level significantly reduced the strength of the material. Therefore, the MCPM level in the current study was reduced to 5–10 wt %. The reduction in MCPM level showed acceptable results, as the increase in MCPM from 5 to 10 wt % showed a negligible effect on the biaxial flexural strength of the material. The level of nisin used in experimental dual-cured composites in a pilot study (4–8 wt %) [12] showed no antibacterial actions. Hence, the level of nisin in the current study was increased to 5–10 wt %. The limitation of the current study was that the nisin powder contains ~90 wt % of NaCl. Hence, both MCPM and nisin fillers may significantly promote water sorption that could plasticize and reduce the rigidity of the polymer network of the adhesives [12,57]. This may, consequently, reduce the strength and modulus of elasticity of the materials.

The effect of nisin on strength reduction was more significant than that of MCPM. This could be due to NaCl in the nisin filler. Additionally, the dissolution of the nisin filler may also leave voids or defects inside the material, which could act as a crack initiator, reducing the strength of the materials. It was proposed that MCPM may be disproportionated and reprecipitated with a different phase of calcium phosphates [58], such as dicalcium phosphates [59]. The new calcium phosphate precipitation may help fill the voids and prevent crack propagation and failure of the composites [10]. Future work should focus on the long-term mechanical properties (>6 months) to ensure the sufficient durability of the materials.

4.3. Water Sorption and Solubility

Resin-based materials can absorb water into the polymer network upon exposure to the oral fluids. Water may plasticize and reduce the rigidity of the polymer networks or cause hydrolytic degradation of the materials [57]. However, it was demonstrated that the absorption of water could promote hygroscopic expansion, which may subsequently help to relieve the polymerization shrinkage stress of the materials [60,61]. Additionally, water is also essential to enable the release of reactive components from rigid resin-based materials.

The maximum level of water sorption and water solubility of the resin-based materials indicated by the BS ISO 4049 was 40 $\mu g/mm^3$ and 7.5 $\mu g/mm^3$, respectively [27]. This indicated that the experimental adhesive with no additives (M0N0) would pass the standard. The additives enhanced water sorption from 22.2 to 233.9 $\mu g/mm^3$ and water solubility from 0.8 to 217.6 $\mu g/mm^3$, due primarily to the hydrophilicity of MCPM and nisin, as was expected. The effect of increasing nisin on water sorption and water solubility was greater than the effect from increasing MCPM. This could be due to the high level of NaCl contained in the nisin powder.

Water solubility may be associated with the loss of components from the materials upon water immersion. Increase in nisin concentration exhibited a greater effect on the water solubility of the adhesives compared with the increase in MCPM concentration. This could be due to the high solubility of nisin and NaCl. Additionally, MCPM may react with water and reprecipitate as calcium phosphate apatite in the materials, which may compensate for the weight loss of the specimens [9]. It can be speculated that the reduction of mass could also be due to leaching out of the unreacted monomers or degradation of the polymer network. It was demonstrated that ester groups in methacrylate monomers were susceptible to hydrolytic and enzymatic degradation [62]. The excessive water dissolution could compromise the physical and mechanical properties of the materials. Hence, the level of MCPM and nisin should be reduced and optimized in future work to ensure that the materials would pass the standard.

4.4. Enamel Shear Bond Strength (SBS)

The sufficient bond strength of the orthodontic adhesives to enamel is crucial to ensure the retention of brackets, allowing the transferred forces from the archwire for orthodontic movement. Additionally, high bond strength may help ensure that the brackets can withstand long-term masticatory forces during the long period of orthodontic treatment, which can be up to 3–5 years [63,64].

The SBS of experimental adhesives in the current study was within the acceptable range reported in the published studies (8–40 MPa) [65,66]. It should be mentioned that the minimum SBS to the enamel of the orthodontic adhesives is not yet specified by the ISO standard. However, the minimum clinically acceptable SBS is ~8 MPa [63]. The addition of hydrophilic additives reduced the SBS of the experimental material, as was expected. The main reason could be the lack of silanation and the increase in water sorption by MCPM and nisin. The negative effect on SBS of the increase in nisin was more evident than the effect from increasing MCPM. This could be due to NaCl contained in nisin promoting excessive water sorption, thereby reducing the strength of adhesives.

The adhesive remnant index (ARI score) is commonly used to determine the remaining adhesive on the tooth surface after debonding. In general, a higher ARI score indicates strong interaction or high bond strength between the adhesive and the enamel surface [67]. The concern of a high ARI score is that excessive bond strength of material to the enamel may lead to the destruction of the enamel surface during debonding [68]. This may cause esthetic problems and increase plaque retention, thereby increasing the risk of caries formation.

A minimum requirement of the ARI score has not yet been specified in an ISO standard. It was expected that the adhesives that exhibited an ARI score of 0 or 1 may provide sufficient bond strength with a low risk of enamel fracture during detachment of brackets [69]. However, this may require additional clinical time to remove the adhesive and clean the enamel surfaces. The most common ARI score of experimental orthodontic adhesive in the current study was 1. Additionally, no enamel breakage was detected in the specimens. This may suggest a low risk of enamel breakdown during debonding in clinical applications. The ARI scores of the experimental adhesives were also in accordance with those observed for commercial orthodontic adhesives in published studies [65,68]. Future work should assess the SBS using long-term thermocycling (5000–10,000 cycles) to determine the strength in accelerated ageing. Additionally, it may be interesting to test SBS under acidic challenge to assess the buffering effects of MCPM on the SBS of the materials.

4.5. Calcium Phosphate Precipitation

Ion-releasing orthodontic adhesives should promote essential ions that could promote mineral precipitation to prevent dental caries [34]. It was suggested that the assessment of the apatite forming ability of the materials in phosphate solution could be a simple method for initial screening the mineralizing ability of the materials [70,71]. The addition of MCPM at 5 or 10 wt % enabled the formation of calcium phosphate precipitation at the interface of orthodontic adhesives in the current study. This finding was consistent with the composites containing a similar level of MCPM in the previously published study [12]. This was beneficial in terms of physical/mechanical properties as the use of a low level of MCPM may not exhibit detrimental effects on the strength of the adhesives. The elemental analysis by EDX demonstrated that the Ca/P ratio of the precipitation was ~1, which may suggest that the mineral precipitation could be dicalcium phosphates (brushite), which is the early phase of hydroxyapatite formation [56]. The test was assessed at an early time (24 h). Hence, future work should determine the calcium phosphate apatite formation at a late time (up to 4 weeks). Additionally, the remineralizing effects on the in vitro demineralized enamel should be assessed to confirm the remineralizing effects of the experimental orthodontic adhesives.

4.6. Ion Release

The ability of materials to release ions was expected to the enhance remineralizing actions of materials [58,72]. Calcium and phosphate ions are essential to encourage the saturated condition for the precipitation of calcium phosphate apatites such as dicalcium phosphate or hydroxyapatite [73]. Additionally, the release of ions also helps buffer the acidic condition during caries attack [74,75]. However, the minimum level of Ca and P ions that can provide clinical remineralizing effects are not concluded.

MCPM contains a low Ca/P ratio (0.5), which suggests that the material can be readily reacted with water and release calcium and phosphate ions. The addition of MCPM in the experimental orthodontic adhesives resulted in the release of Ca and P ions, which was in agreement with the result in a published study [76]. The increase in MCPM level from 5 to 10 wt % enhanced the Ca and P release, as was expected. The increase in nisin also promoted the release of Ca and P ions. A possible mechanism could be that nisin encouraged water sorption, thus enhancing the dissolution of MCPM. A limitation of the current study is that the measurement was not performed at different time intervals, so determining the release over time was not possible. Future works should assess the ions released at different time points, or their release in acidic conditions to mimic the cariogenic challenge.

4.7. Antibacterial Action on *S. mutans*

The colonization of biofilm around the excess orthodontic adhesives could increase the risk of developing white spot lesions around the brackets [77,78]. Meticulous oral hygiene and various additional oral health care products [79,80] are essential to help control plaque accumulation during fixed orthodontic treatments. However, the success of the interventions relies upon the favorable compliance of patients. The aim of adding antibacterial agents into orthodontic adhesives was, therefore, to help inhibit the bacterial growth, which could subsequently reduce the risk of demineralization [81].

The limitation of the current study showed that the addition of nisin into the experimental orthodontic adhesives failed to demonstrate the inhibitory effects on *S. mutans*. A possible explanation is that the concentration of nisin used in the current study (5–10 wt %) was too low to exhibit the significant benefit. It should be mentioned that previous studies showed that the addition of nisin in commercial dentin bonding agents for ~3–5 wt % promoted the antibacterial action of the materials [24,25]. The dentin bonding agents usually contain a high level of low-molecular-weight and hydrophilic monomers. The highly flexible polymer network in the dentin bonding agent may therefore encourage the diffusion and release of nisin from the material. Although the current study incorporated a higher level of nisin, the rigid polymer network of orthodontic adhesive may limit the diffusion and release of nisin. A study indicated that the concentration of nisin that can inhibit the growth of *S. mutans* was 10 µg/mL [82]. A limitation of the current study was that the concentration of released nisin was not analyzed. Hence, the release kinetic of nisin using HPLC should be included in future work. This would help to optimize the required concentration of nisin. Additionally, ultrapure nisin (concentration of nisin

>95 wt %) should be used in future work to increase the concentration of nisin in the materials.

5. Conclusions

Ion-releasing and antibacterial experimental orthodontic adhesives containing MCPM and nisin were prepared. The additives showed minimal effect on the degree of monomer conversion but reduced the mechanical properties of the materials. However, the strength was still within the acceptable level required by the ISO standard. The additives also increased the water sorption/solubility of the materials. The addition of Nisin demonstrated no inhibition effect on the growth of *S. mutans*. The addition of MCPM promoted ion release and calcium phosphate precipitation for the adhesive. This was expected to promote the remineralizing properties of the materials.

Author Contributions: Conceptualization, K.I., S.B., V.A. and P.P.; methodology, S.C., W.C., K.I., S.B., V.A. and P.P.; software, S.C., W.C. and P.P.; validation, S.C., W.C. and P.P.; formal analysis, S.C., W.C. and P.P.; investigation, S.C. and W.C.; resources, V.A. and P.P.; data curation, S.C. and W.C.; writing—original draft preparation, S.C. and P.P.; writing—review and editing, S.C. and P.P.; visualization, S.C. and P.P.; supervision, K.I., S.B., V.A. and P.P.; project administration, P.P.; funding acquisition, K.I., S.B., V.A. and P.P. All authors have read and agreed to the published version of the manuscript.

Funding: This research was supported by Faculty of Dentistry, Thammasat University.

Institutional Review Board Statement: The study was conducted according to the guidelines of the Declaration of Helsinki, and approved by the Ethics Review Subcommittee Board for Human Research Involving Sciences, Thammasat University, no. 3 (Faculty of Health Sciences and Science and Technology, Thammasat University, protocol code 152/2563, date of approval: 11 November 2020).

Informed Consent Statement: Patient consent was waived because the identification of the extracted teeth was not required.

Data Availability Statement: The datasets generated and/or analyzed during the current study are available from the corresponding author on reasonable request.

Acknowledgments: The authors are grateful for the support from the Thammasat University Research Unit in Dental and Bone Substitute Biomaterials, Thammasat University. The syringes for the experimental orthodontic adhesives were provided by Sulzer Mixpac AG (Switzerland). The authors would like to thank Arnit Toneluck at Medicinal Extracts and Biomaterials Laboratory at the Faculty of Dentistry Thammasat University for technical support.

Conflicts of Interest: The authors declare no conflict of interest. The funders had no role in the design of the study; in the collection, analyses, or interpretation of data; in the writing of the manuscript, or in the decision to publish the results.

References

1. Sonesson, M.; Brechter, A.; Lindman, R.; Abdulraheem, S.; Twetman, S. Fluoride varnish for white spot lesion prevention during orthodontic treatment: Results of a randomized controlled trial 1 year after debonding. *Eur. J. Orthod.* **2021**, *43*, 473–477. [CrossRef]
2. Livas, C.; Kuijpers-Jagtman, A.M.; Bronkhorst, E.; Derks, A.; Katsaros, C. Quantification of White Spot Lesions around Orthodontic Brackets with Image Analysis. *Angle Orthod.* **2008**, *78*, 585–590. [CrossRef]
3. Dalessandri, D.; Dalessandri, M.; Bonetti, S.; Visconti, L.; Paganelli, C. Effectiveness of an indirect bonding technique in reducing plaque accumulation around braces. *Angle Orthod.* **2012**, *82*, 313–318. [CrossRef]
4. Almosa, N.A.; Alqasir, A.M.; Aldekhayyil, M.A.; Aljelayel, A.; Aldosari, M.A. Enamel demineralization around two different orthodontic bracket adhesive systems: An in vivo study. *Saudi Dent. J.* **2019**, *31*, 99–104. [CrossRef]
5. Al Tuma, R.R.; Yassir, Y.A. Evaluation of a newly developed calcium fluoride nanoparticles-containing orthodontic primer: An in-vitro study. *J. Mech. Behav. Biomed. Mater.* **2021**, *122*, 104691. [CrossRef]
6. Al-Eesa, N.A.; Johal, A.; Hill, R.G.; Wong, F.S.L. Fluoride containing bioactive glass composite for orthodontic adhesives—Apatite formation properties. *Dent. Mater.* **2018**, *34*, 1127–1133. [CrossRef]
7. Liu, Y.; Zhang, L.; Niu, L.N.; Yu, T.; Xu, H.H.K.; Weir, M.D.; Oates, T.W.; Tay, F.R.; Chen, J.H. Antibacterial and remineralizing orthodontic adhesive containing quaternary ammonium resin monomer and amorphous calcium phosphate nanoparticles. *J. Dent.* **2018**, *72*, 53–63. [CrossRef]

8. Stammeier, J.A.; Purgstaller, B.; Hippler, D.; Mavromatis, V.; Dietzel, M. In-situ Raman spectroscopy of amorphous calcium phosphate to crystalline hydroxyapatite transformation. *MethodsX* **2018**, *5*, 1241–1250. [CrossRef]
9. Kangwankai, K.; Sani, S.; Panpisut, P.; Xia, W.; Ashley, P.; Petridis, H.; Young, A.M. Monomer conversion, dimensional stability, strength, modulus, surface apatite precipitation and wear of novel, reactive calcium phosphate and polylysine-containing dental composites. *PLoS ONE* **2017**, *12*, e0187757. [CrossRef]
10. Panpisut, P.; Khan, M.A.; Main, K.; Arshad, M.; Xia, W.; Petridis, H.; Young, A.M. Polymerization kinetics stability, volumetric changes, apatite precipitation, strontium release and fatigue of novel bone composites for vertebroplasty. *PLoS ONE* **2019**, *14*, e0207965. [CrossRef]
11. Panpisut, P.; Liaqat, S.; Zacharaki, E.; Xia, W.; Petridis, H.; Young, A.M. Dental composites with calcium/strontium phosphates and polylysine. *PLoS ONE* **2016**, *11*, e0164653. [CrossRef]
12. Panpisut, P.; Suppapatpong, T.; Rattanapan, A.; Wongwarawut, P. Monomer conversion, biaxial flexural strength, apatite forming ability of experimental dual-cured and self-adhesive dental composites containing calcium phosphate and nisin. *Dent. Mater. J.* **2021**, *40*, 399–406. [CrossRef]
13. Aljabo, A.; Abou Neel, E.A.; Knowles, J.C.; Young, A.M. Development of dental composites with reactive fillers that promote precipitation of antibacterial-hydroxyapatite layers. *Mater. Sci. Eng. C Mater. Biol. Appl.* **2016**, *60*, 285–292. [CrossRef]
14. Dorozhkin, S.V. Calcium orthophosphates as a dental regenerative material. In *Advanced Dental Biomaterials*; Elsevier: Duxford, UK, 2019; pp. 377–452.
15. Aljabo, A.; Xia, W.; Liaqat, S.; Khan, M.A.; Knowles, J.C.; Ashley, P.; Young, A.M. Conversion, shrinkage, water sorption, flexural strength and modulus of re-mineralizing dental composites. *Dent. Mater.* **2015**, *31*, 1279–1289. [CrossRef]
16. Araujo, J.; Alvim, M.M.A.; Campos, M.; Apolonio, A.C.M.; Carvalho, F.G.; Lacerda-Santos, R. Analysis of Chlorhexidine Modified Cement in Orthodontic Patients: A Double-Blinded, Randomized, Controlled Trial. *Eur. J. Dent.* **2021**, *15*, 639–646. [CrossRef]
17. Opstrup, M.S.; Jemec, G.B.E.; Garvey, L.H. Chlorhexidine Allergy: On the Rise and Often Overlooked. *Curr. Allergy Asthma Rep.* **2019**, *19*, 23. [CrossRef]
18. Pemberton, M.N. Allergy to Chlorhexidine. *Dent. Update* **2016**, *43*, 272–274. [CrossRef]
19. Saleem, H.G.; Seers, C.A.; Sabri, A.N.; Reynolds, E.C. Dental plaque bacteria with reduced susceptibility to chlorhexidine are multidrug resistant. *BMC Microbiol.* **2016**, *16*, 214. [CrossRef]
20. Cieplik, F.; Jakubovics, N.S.; Buchalla, W.; Maisch, T.; Hellwig, E.; Al-Ahmad, A. Resistance Toward Chlorhexidine in Oral Bacteria—Is There Cause for Concern? *Front. Microbiol.* **2019**, *10*, 587. [CrossRef]
21. Webber, J.L.; Namivandi-Zangeneh, R.; Drozdek, S.; Wilk, K.A.; Boyer, C.; Wong, E.H.H.; Bradshaw-Hajek, B.H.; Krasowska, M.; Beattie, D.A. Incorporation and antimicrobial activity of nisin Z within carrageenan/chitosan multilayers. *Sci. Rep.* **2021**, *11*, 1690. [CrossRef]
22. Zhou, H.; Fang, J.; Tian, Y.; Lu, X.Y. Mechanisms of nisin resistance in Gram-positive bacteria. *Ann. Microbiol.* **2014**, *64*, 413–420. [CrossRef]
23. Prince, A.; Sandhu, P.; Ror, P.; Dash, E.; Sharma, S.; Arakha, M.; Jha, S.; Akhter, Y.; Saleem, M. Lipid-II Independent Antimicrobial Mechanism of Nisin Depends on Its Crowding and Degree of Oligomerization. *Sci. Rep.* **2016**, *6*, 37908. [CrossRef]
24. Zhao, M.; Qu, Y.; Liu, J.; Mai, S.; Gu, L. A universal adhesive incorporating antimicrobial peptide nisin: Effects on Streptococcus mutans and saliva-derived multispecies biofilms. *Odontology* **2020**, *108*, 376–385. [CrossRef]
25. Su, M.; Yao, S.; Gu, L.; Huang, Z.; Mai, S. Antibacterial effect and bond strength of a modified dental adhesive containing the peptide nisin. *Peptides* **2018**, *99*, 189–194. [CrossRef]
26. Delgado, A.H.S.; Young, A.M. Methacrylate peak determination and selection recommendations using ATR-FTIR to investigate polymerisation of dental methacrylate mixtures. *PLoS ONE* **2021**, *16*, e0252999. [CrossRef]
27. British Standard. BS EN ISO 4049:2019. In *Dentistry-Polymer-Based Restorative Materials*; BSI Standards Limited: London, UK, 2019.
28. British Standard. PD ISO/TS 11405:2015. In *Dentistry-Testing of Adhesion to Tooth Structure*; BSI Standards Limited: London, UK, 2015.
29. Thepveera, W.; Potiprapanpong, W.; Toneluck, A.; Channasanon, S.; Khamsuk, C.; Monmaturapoj, N.; Tanodekaew, S.; Panpisut, P. Rheological Properties, Surface Microhardness, and Dentin Shear Bond Strength of Resin-Modified Glass Ionomer Cements Containing Methacrylate-Functionalized Polyacids and Spherical Pre-Reacted Glass Fillers. *J. Funct. Biomater.* **2021**, *12*, 42. [CrossRef]
30. Årtun, J.; Bergland, S. Clinical trials with crystal growth conditioning as an alternative to acid-etch enamel pretreatment. *Am. J. Orthod.* **1984**, *85*, 333–340. [CrossRef]
31. Gonzalez-Serrano, C.; Baena, E.; Fuentes, M.V.; Albaladejo, A.; Miguez-Contreras, M.; Lagravere, M.O.; Ceballos, L. Shear bond strength of a flash-free orthodontic adhesive system after thermal aging procedure. *J. Clin. Exp. Dent.* **2019**, *11*, e154–e161. [CrossRef]
32. Potiprapanpong, W.; Thepveera, W.; Khamsuk, C.; Channasanon, S.; Tanodekaew, S.; Patntirapong, S.; Monmaturapoj, N.; Panpisut, P. Monomer Conversion, Dimensional Stability, Biaxial Flexural Strength, Ion Release, and Cytotoxicity of Resin-Modified Glass Ionomer Cements Containing Methacrylate-Functionalized Polyacids and Spherical Pre-Reacted Glass Fillers. *Polymers* **2021**, *13*, 2742. [CrossRef]
33. Lygidakis, N.N.; Allan, E.; Xia, W.; Ashley, P.F.; Young, A.M. Early Polylysine Release from Dental Composites and Its Effects on Planktonic Streptococcus mutans Growth. *J. Funct. Biomater.* **2020**, *11*, 53. [CrossRef]

34. Ferreira, C.J.; Leitune, V.C.B.; Balbinot, G.S.; Degrazia, F.W.; Arakelyan, M.; Sauro, S.; Mezzomo Collares, F. Antibacterial and Remineralizing Fillers in Experimental Orthodontic Adhesives. *Materials* **2019**, *12*, 652. [CrossRef]
35. Panpisut, P.; Toneluck, A. Monomer conversion, dimensional stability, biaxial flexural strength, and fluoride release of resin-based restorative material containing alkaline fillers. *Dent. Mater. J.* **2020**, *39*, 608–615. [CrossRef]
36. Putzeys, E.; Nys, S.; Cokic, S.M.; Duca, R.C.; Vanoirbeek, J.; Godderis, L.; Meerbeek, B.V.; Van Landuyt, K.L. Long-term elution of monomers from resin-based dental composites. *Dent. Mater.* **2019**, *35*, 477–485. [CrossRef]
37. Bationo, R.; Rouamba, A.; Diarra, A.; Beugre-Kouassi, M.L.A.; Beugre, J.B.; Jordana, F. Cytotoxicity evaluation of dental and orthodontic light-cured composite resins. *Clin. Exp. Dent. Res.* **2021**, *7*, 40–48. [CrossRef]
38. Boonen, I.; De Nys, S.; Vervliet, P.; Covaci, A.; Van Landuyt, K.L.; Duca, R.C.; Godderis, L.; Denison, M.S.; Elskens, M. Assessing the estrogenic activity of chemicals present in resin based dental composites and in leachates of commercially available composites using the ERalpha-CALUX bioassay. *Dent. Mater.* **2021**, *37*, 1834–1844. [CrossRef]
39. Kim, K.; An, J.S.; Lim, B.S.; Ahn, S.J. Effect of Bisphenol A Glycol Methacrylate on Virulent Properties of Streptococcus mutans UA159. *Caries Res.* **2019**, *53*, 84–95. [CrossRef] [PubMed]
40. Kim, K.; Kim, J.N.; Lim, B.S.; Ahn, S.J. Urethane Dimethacrylate Influences the Cariogenic Properties of Streptococcus Mutans. *Materials* **2021**, *14*, 1015. [CrossRef]
41. Yilmaz, B.; Bakkal, M.; Zengin Kurt, B. Structural and mechanical analysis of three orthodontic adhesive composites cured with different light units. *J. Appl. Biomater. Funct. Mater.* **2020**, *18*, 2280800020901716. [CrossRef]
42. Lempel, E.; Ori, Z.; Kincses, D.; Lovasz, B.V.; Kunsagi-Mate, S.; Szalma, J. Degree of conversion and in vitro temperature rise of pulp chamber during polymerization of flowable and sculptable conventional, bulk-fill and short-fibre reinforced resin composites. *Dent. Mater.* **2021**, *37*, 983–997. [CrossRef]
43. Walters, N.J.; Xia, W.; Salih, V.; Ashley, P.F.; Young, A.M. Poly(propylene glycol) and urethane dimethacrylates improve conversion of dental composites and reveal complexity of cytocompatibility testing. *Dent. Mater.* **2016**, *32*, 264–277. [CrossRef]
44. Sideridou, I.; Tserki, V.; Papanastasiou, G. Effect of chemical structure on degree of conversion in light-cured dimethacrylate-based dental resins. *Biomaterials* **2002**, *23*, 1819–1829. [CrossRef]
45. Shortall, A.C.; Palin, W.M.; Burtscher, P. Refractive index mismatch and monomer reactivity influence composite curing depth. *J. Dent. Res.* **2008**, *87*, 84–88. [CrossRef]
46. Fujita, K.; Nishiyama, N.; Nemoto, K.; Okada, T.; Ikemi, T. Effect of base monomer's refractive index on curing depth and polymerization conversion of photo-cured resin composites. *Dent. Mater. J.* **2005**, *24*, 403–408. [CrossRef]
47. Kumar, N.; Ghani, F.; Fareed, M.A.; Riaz, S.; Khurshid, Z.; Zafar, M.S. Bi-axial flexural strength of resin based dental composites—Influence and reliability of the testing method configuration. *Mater. Technol.* **2021**, 1–7. [CrossRef]
48. Ilie, N.; Hilton, T.; Heintze, S.; Hickel, R.; Watts, D.; Silikas, N.; Stansbury, J.; Cadenaro, M.; Ferracane, J. Academy of dental materials guidance—Resin composites: Part I—Mechanical properties. *Dent. Mater.* **2017**, *33*, 880–894. [CrossRef]
49. Miura, D.; Ishida, Y.; Miyasaka, T.; Aoki, H.; Shinya, A. Reliability of Different Bending Test Methods for Dental Press Ceramics. *Materials* **2020**, *13*, 5162. [CrossRef] [PubMed]
50. Azad, E.; Atai, M.; Zandi, M.; Shokrollahi, P.; Solhi, L. Structure-properties relationships in dental adhesives: Effect of initiator, matrix monomer structure, and nano-filler incorporation. *Dent. Mater.* **2018**, *34*, 1263–1270. [CrossRef] [PubMed]
51. Gomes de Araujo-Neto, V.; Sebold, M.; Fernandes de Castro, E.; Feitosa, V.P.; Giannini, M. Evaluation of physico-mechanical properties and filler particles characterization of conventional, bulk-fill, and bioactive resin-based composites. *J. Mech. Behav. Biomed. Mater.* **2021**, *115*, 104288. [CrossRef]
52. Odermatt, R.; Mohn, D.; Wiedemeier, D.B.; Attin, T.; Tauböck, T.T. Bioactivity and physico-chemical properties of dental composites functionalized with nano-vs. micro-sized bioactive glass. *J. Clin. Med.* **2020**, *9*, 772. [CrossRef]
53. Par, M.; Tarle, Z.; Hickel, R.; Ilie, N. Mechanical properties of experimental composites containing bioactive glass after artificial aging in water and ethanol. *Clin. Oral Investig.* **2019**, *23*, 2733–2741. [CrossRef] [PubMed]
54. Natale, L.C.; Rodrigues, M.C.; Alania, Y.; Chiari, M.D.S.; Boaro, L.C.C.; Cotrim, M.; Vega, O.; Braga, R.R. Mechanical characterization and ion release of bioactive dental composites containing calcium phosphate particles. *J. Mech. Behav. Biomed. Mater.* **2018**, *84*, 161–167. [CrossRef]
55. Xu, H.H.; Moreau, J.L. Dental glass-reinforced composite for caries inhibition: Calcium phosphate ion release and mechanical properties. *J. Biomed. Mater. Res. B Appl. Biomater.* **2010**, *92*, 332–340. [CrossRef]
56. Dorozhkin, S.V. Calcium orthophosphates ($CaPO_4$): Occurrence and properties. *Prog. Biomater.* **2016**, *5*, 9–70. [CrossRef] [PubMed]
57. Ferracane, J.L. Hygroscopic and hydrolytic effects in dental polymer networks. *Dent. Mater.* **2006**, *22*, 211–222. [CrossRef] [PubMed]
58. Braga, R.R. Calcium phosphates as ion-releasing fill.lers in restorative resin-based materials. *Dent. Mater.* **2019**, *35*, 3–14. [CrossRef]
59. Mehdawi, I.M.; Pratten, J.; Spratt, D.A.; Knowles, J.C.; Young, A.M. High strength re-mineralizing, antibacterial dental composites with reactive calcium phosphates. *Dent. Mater.* **2013**, *29*, 473–484. [CrossRef]
60. Suiter, E.A.; Watson, L.E.; Tantbirojn, D.; Lou, J.S.; Versluis, A. Effective Expansion: Balance between Shrinkage and Hygroscopic Expansion. *J. Dent. Res.* **2016**, *95*, 543–549. [CrossRef]

61. Sokolowski, K.; Szczesio-Wlodarczyk, A.; Bociong, K.; Krasowski, M.; Fronczek-Wojciechowska, M.; Domarecka, M.; Sokolowski, J.; Lukomska-Szymanska, M. Contraction and Hydroscopic Expansion Stress of Dental Ion-Releasing Polymeric Materials. *Polymers* **2018**, *10*, 1093. [CrossRef] [PubMed]
62. Podgórski, M.; Becka, E.; Claudino, M.; Flores, A.; Shah, P.K.; Stansbury, J.W.; Bowman, C.N. Ester-free thiol–ene dental restoratives—Part A: Resin development. *Dent. Mater.* **2015**, *31*, 1255–1262. [CrossRef] [PubMed]
63. Reynolds, I. A review of direct orthodontic bonding. *Br. J. Orthod.* **1975**, *2*, 171–178. [CrossRef]
64. Bakhadher, W.; Halawany, H.; Talic, N.; Abraham, N.; Jacob, V. Factors Affecting the Shear Bond Strength of Orthodontic Brackets—A Review of In Vitro Studies. *Acta Medica* **2015**, *58*, 43–48. [CrossRef]
65. Mitwally, R.A.; Bakhsh, Z.T.; Feteih, R.M.; Bakry, A.S.; Abbassy, M.A. Orthodontic Bracket Bonding Using Self-adhesive Cement to Facilitate Bracket Debonding. *J. Adhes. Dent.* **2019**, *21*, 551–556. [CrossRef]
66. Eslamian, L.; Borzabadi-Farahani, A.; Karimi, S.; Saadat, S.; Badiee, M.R. Evaluation of the Shear Bond Strength and Antibacterial Activity of Orthodontic Adhesive Containing Silver Nanoparticle, an In-Vitro Study. *Nanomaterials* **2020**, *10*, 1466. [CrossRef]
67. Henkin, F.D.S.; de Oliveira Dias de Macêdo, É.; Santos, K.D.S.; Schwarzbach, M.; Samuel, S.M.W.; Mundstock, K.S. In vitro analysis of shear bond strength and adhesive remnant index of different metal brackets. *Dent. Press J. Orthod.* **2016**, *21*, 67–73. [CrossRef] [PubMed]
68. Sharma, S.; Tandon, P.; Nagar, A.; Singh, G.P.; Singh, A.; Chugh, V.K. A comparison of shear bond strength of orthodontic brackets bonded with four different orthodontic adhesives. *J. Orthod. Sci.* **2014**, *3*, 29–33. [CrossRef] [PubMed]
69. Vinagre, A.R.; Messias, A.L.; Gomes, M.A.; Costa, A.L.; Ramos, J.C. Effect of time on shear bond strength of four orthodontic adhesive systems. *J. Rev. Port. Estomatol. Med. Dentária Cir. Maxilofac.* **2014**, *55*, 142–151. [CrossRef]
70. Gandolfi, M.G.; Taddei, P.; Siboni, F.; Modena, E.; De Stefano, E.D.; Prati, C. Biomimetic remineralization of human dentin using promising innovative calcium-silicate hybrid "smart" materials. *Dent. Mater.* **2011**, *27*, 1055–1069. [CrossRef] [PubMed]
71. Khan, A.S.; Syed, M.R. A review of bioceramics-based dental restorative materials. *Dent. Mater. J.* **2019**, *38*, 163–176. [CrossRef]
72. Yi, J.; Dai, Q.; Weir, M.D.; Melo, M.A.S.; Lynch, C.D.; Oates, T.W.; Zhang, K.; Zhao, Z.; Xu, H.H.K. A nano-CaF2-containing orthodontic cement with antibacterial and remineralization capabilities to combat enamel white spot lesions. *J. Dent.* **2019**, *89*, 103172. [CrossRef]
73. Aoba, T. Solubility properties of human tooth mineral and pathogenesis of dental caries. *Oral Dis.* **2004**, *10*, 249–257. [CrossRef] [PubMed]
74. Wanitwisutchai, T.; Monmaturapoj, N.; Srisatjaluk, R.; Subannajui, K.; Dechkunakorn, S.; Anuwongnukroh, N.; Pongprueksa, P. Buffering capacity and antibacterial properties among bioactive glass-containing orthodontic adhesives. *Dent. Mater. J.* **2021**, 2020–2375. [CrossRef]
75. Cieplik, F.; Rupp, C.M.; Hirsch, S.; Muehler, D.; Enax, J.; Meyer, F.; Hiller, K.-A.; Buchalla, W. Ca^{2+} release and buffering effects of synthetic hydroxyapatite following bacterial acid challenge. *BMC Oral Health* **2020**, *20*, 85. [CrossRef]
76. Mehdawi, I.; Neel, E.A.; Valappil, S.P.; Palmer, G.; Salih, V.; Pratten, J.; Spratt, D.A.; Young, A.M. Development of remineralizing, antibacterial dental materials. *Acta Biomater.* **2009**, *5*, 2525–2539. [CrossRef]
77. Sukontapatipark, W.; el-Agroudi, M.A.; Selliseth, N.J.; Thunold, K.; Selvig, K.A. Bacterial colonization associated with fixed orthodontic appliances. A scanning electron microscopy study. *Eur. J. Orthod.* **2001**, *23*, 475–484. [CrossRef] [PubMed]
78. Condò, R.; Mampieri, G.; Pasquantonio, G.; Giancotti, A.; Pirelli, P.; Cataldi, M.E.; La Rocca, S.; Leggeri, A.; Notargiacomo, A.; Maiolo, L. In vitro evaluation of structural factors favouring bacterial adhesion on orthodontic adhesive resins. *J. Mater. Sci.* **2021**, *14*, 2485. [CrossRef]
79. Hadj-Hamou, R.; Senok, A.C.; Athanasiou, A.E.; Kaklamanos, E.G. Do probiotics promote oral health during orthodontic treatment with fixed appliances? A systematic review. *BMC Oral Health* **2020**, *20*, 126. [CrossRef]
80. Nardi, G.M.; Fais, S.; Casu, C.; Mazur, M.; Di Giorgio, R.; Grassi, R.; Grassi, F.R.; Orru, G. Mouthwash Based on Ozonated Olive Oil in Caries Prevention: A Preliminary In-Vitro Study. *Int J. Environ. Res. Public Health* **2020**, *17*, 9106. [CrossRef]
81. de Almeida, C.M.; da Rosa, W.L.O.; Meereis, C.T.W.; de Almeida, S.M.; Ribeiro, J.S.; da Silva, A.F.; Lund, R.G. Efficacy of antimicrobial agents incorporated in orthodontic bonding systems: A systematic review and meta-analysis. *J. Orthod.* **2018**, *45*, 79–93. [CrossRef] [PubMed]
82. Shin, J.M.; Ateia, I.; Paulus, J.R.; Liu, H.; Fenno, J.C.; Rickard, A.H.; Kapila, Y.L. Antimicrobial nisin acts against saliva derived multi-species biofilms without cytotoxicity to human oral cells. *Front. Microbiol.* **2015**, *6*, 617. [CrossRef] [PubMed]

Journal of
Functional Biomaterials

MDPI

Review

Properties and Applications of PDMS for Biomedical Engineering: A Review

Inês Miranda [1], Andrews Souza [2], Paulo Sousa [1], João Ribeiro [3], Elisabete M. S. Castanheira [4], Rui Lima [2,5,*] and Graça Minas [1]

1 Center for MicroElectromechanical Systems (CMEMS-UMinho), Campus de Azurém, University of Minho, 4800-058 Guimaraes, Portugal; ines_sofia_miranda@outlook.com (I.M.); psousa@dei.uminho.pt (P.S.); gminas@dei.uminho.pt (G.M.)
2 MEtRICs, Mechanical Engineering Department, Campus de Azurém, University of Minho, 4800-058 Guimaraes, Portugal; andrewsv81@gmail.com
3 Centro de Investigação de Montanha (CIMO), Campus de Santa Apolónia, Instituto Politécnico de Bragança, 5300-253 Braganca, Portugal; jribeiro@ipb.pt
4 Centre of Physics of Minho and Porto Universities (CF-UM-UP), Campus de Gualtar, University of Minho, 4710-057 Braga, Portugal; ecoutinho@fisica.uminho.pt
5 CEFT, Faculdade de Engenharia da Universidade do Porto (FEUP), Rua Roberto Frias, 4200-465 Porto, Portugal
* Correspondence: rl@dem.uminho.pt

Abstract: Polydimethylsiloxane (PDMS) is an elastomer with excellent optical, electrical and mechanical properties, which makes it well-suited for several engineering applications. Due to its biocompatibility, PDMS is widely used for biomedical purposes. This widespread use has also led to the massification of the soft-lithography technique, introduced for facilitating the rapid prototyping of micro and nanostructures using elastomeric materials, most notably PDMS. This technique has allowed advances in microfluidic, electronic and biomedical fields. In this review, an overview of the properties of PDMS and some of its commonly used treatments, aiming at the suitability to those fields' needs, are presented. Applications such as microchips in the biomedical field, replication of cardiovascular flow and medical implants are also reviewed.

Keywords: polydimethylsiloxane; PDMS properties; PDMS applications; microfluidics; biomedical engineering

1. Introduction

Polydimethylsiloxane (PDMS) is an elastomeric polymer with interesting properties for biomedical applications, including physiological indifference, excellent resistance to biodegradation, biocompatibility, chemical stability, gas permeability, good mechanical properties, excellent optical transparency and simple fabrication by replica moulding [1–5]. Due to these characteristics, PDMS has been widely used in micropumps [6], catheter surfaces [7], dressings and bandages [8], microvalves [9], optical systems [10,11], in the in vitro study of diseases [12,13], in implants [14,15], in microfluidics and photonics [16–19]. Moreover, soft-lithography technology has driven the use of PDMS in microelectromechanical systems (MEMS) applications and in microfluidic components [17,18,20]. Soft-lithography techniques such as micro-contact printing, replica moulding, micro-transfer moulding, micro-moulding in capillaries and solvent-assisted micro-moulding usually require the use of PDMS to create an elastomeric stamp or mould that incorporates nano- and microstructures for the transfer of patterns onto a subsequent substrate [18,21].

MEMS are approaches that use electronic and mechanical technologies to deal with biomedical problems on the micro-scale [22]. MEMS-based devices have been widely used in the biomedical area for applications such as diagnostics and therapeutics. These systems can be microsensors or microtransducers, and are helpful in areas such as physics, mechanics, electronics and biomedicals, as they can provide very precise and fast results [23].

The investigation and improvement of already existing MEMS are more and more common. As they are increasingly commercialized, the necessity to find processes and materials that enable mass production while reducing cost has emerged [21]. MEMS are traditionally silicon-based and the pursuit for a more biologically friendly material is needed. Polymers allow rapid prototyping and mass production techniques as well as having a lower cost in relation to silicon, making them particularly attractive for the development of MEMS [21]. Photolithography is the most commonly used technique in microfabrication, however, this method is expensive [24]. With the introduction of polymers in microsystems, new manufacturing techniques have been studied, such as soft-lithography, which can be a cheaper method comparatively to photolithography, even when a costly mould is needed for patterning; once a mould is created, it can be reused several times [20]. Additionally, there are alternatives which are attempting to reduce the cost of the moulds, relying on cleanroom less approaches [25]. Candidate polymers for the production of MEMS are polycarbonate (PC), polymethylmethacrylate (PMMA), polyvinylchloride (PVC), polyethylene (PE) and PDMS [21].

Additionally, PDMS is the most commonly used material in the manufacturing of microfluidic devices, which are an important technology for the development of systems such as drug delivery, DNA sequencing, clinical diagnostics, point of care testing and chemical synthesis [26]. The used materials in these systems should be biocompatible, optically transparent and provide fast prototyping and low fabrication cost [27], features found in PDMS.

In addition to applications in microfluidics, PDMS has been widely used in the fabrication of biomodels (flow phantom) for the in vitro hemodynamic study of diseases such as aneurysms and stenosis [28–31]. The biomodels developed in PDMS allow good replicability of the lumen of the arteries and good transparency, being ideal for the application of optical techniques of micro particle image velocimetry (micro-PIV), particle image velocimetry (PIV), particle tracking velocimetry (PTV) and non-evasive techniques [32–34]. These experimental tests have provided a greater understanding of these pathologies, validated numerical techniques and tested medical devices such as stents [35–37].

PDMS has also been investigated in the field of medical implants [38–42]. These types of implants are usually made with titanium or its alloys; however, such materials do not allow a good osseointegration [39]. In order to overcome this limitation, PDMS has been studied to produce coatings with microscale features that help the bonding between the implant and the bone. The main characteristics for its use in implants are its high biocompatibility, excellent resistance to biodegradation and flexibility, which makes PDMS one of the most successful polymers in implanted devices, presenting only mild foreign body reactions [43–45]. Common applications include cardiac pacemakers, cuff and book electrodes in the PNS, cochlear implants, bladder and pain controllers and planar electrode arrays in the CNS [45,46].

In this review, research on PDMS properties, their fabrication processes and their characterization methods are reported. Moreover, their use in MEMS applications, microfluidics, medical implants and hemodynamic studies is investigated. Written in a concise, but complete manner, we believe that this manuscript joins together the main advantages, disadvantages and challenges of PMDS when biomedical applications are needed and, therefore, can be extremely useful for researchers looking to learn about this biomaterial and its applicability in this biomedical field.

2. PDMS Properties

Silicon, glass and polymers are the typical materials used for micro devices fabrication: silicon, because of its thermal conductivity and the availability of advanced fabrication technologies; glass, mainly due to its transparency; polymers, because of its low cost, optical transparency and flexibility. Compared to glass and silicon, PDMS turns out to be the most promising elastomer, because the other two materials have a high manufacturing cost, require greater labour intensity and are rigid in nature. The variable elasticity of PDMS in medical applications is also favourable; its modulus of elasticity is 1–3 MPa (compared to ~50 GPa of glass) [2,47]. PDMS is also chemically inert, thermally stable, permeable to gases, simple to handle and manipulate, exhibits isotropic and homogeneous properties and can replicate submicron features to develop microstructures [19,21,48]. Additionally, this elastomer is optically transparent, can work as a thermal and electrical insulator and degrades quickly in the natural environment [49]. PDMS presents a hyperelastic behaviour, which is the ability of a material to undergo large deformations before rupture [50]. This characteristic is also found in biological tissues and, for that reason, PDMS is a well-suited material to mimic, for example, blood vessels [49]. Another characteristic of this elastomer is its biocompatibility, which means that PDMS is compatible with biologic tissues [49]. PDMS presents a transmittance up to 90% for the wavelength from 390 nm to 780 nm [51–53] and, due to this characteristic, PDMS-based microsystems allow the direct observation of the mimicked blood flow inside the mimicked vessels and the integration of optical detection systems, hence playing an important role in this field.

With the purpose of extending the lifespan of a chip, PDMS is used to embed or encapsulate electronic components by casting. Due to its thermal and electrical insulation capability, PDMS protects the components from environmental factors and mechanical shock within a large temperature range (−50–200 °C) [23,48]. In Table 1, some physical properties of PDMS are listed.

Table 1. Typical properties of cured PDMS.

Property (Unity)	Result	References
Transmittance at range 390 nm to 780 nm (%)	75–92	[54,55]
Index of refraction	1.4	[56]
Thermal conductivity (W/m·K)	0.2–0.27	[57,58]
Specific heat (kJ/kg·K)	1.46	[56]
Dielectric strength (kV/mm)	19	[57]
Dielectric constant	2.3–2.8	[56]
Electrical conductivity (ohm·m)	4×10^{13}	[56]
Volume resistivity (ohm·cm)	2.9×10^{14}	[57]
Young's modulus [kPa]	360–870	[59]
Poisson ratio	0.5	[60]
Tensile strength (MPa)	2.24–6.7	[56,57]
Hardness [Shore A]	41–43	[55,61]
Viscosity (Pa·s)	3.5	[57]
Hydrophobicity—contact angle (°)	$\sim 108° \pm 7°$	[62]
Melting Point (°C)	−49.9 to −40	[63]

Despite these advantages, PDMS has some properties that can present a limitation in some applications. Due to its CH_3 groups, PDMS presents a hydrophobic surface (contact angle with water $\sim 108° \pm 7°$) [62,64,65], often limiting its application in solutions composed of biological samples [66]. Additionally, PDMS tends to swell when combined with certain reagents [17,48]. In some applications, the absorption of small molecules flowing through the channels makes it difficult to quantitatively analyse experiments in proteomic drug discovery and cell culture [67,68]. In microchannels, the hydrophobicity of PDMS generates complications that include impedance to the flow of polar liquids, which makes it difficult to wet its surface with aqueous solvents [49]. On the other hand, much effort has been made to make the PDMS surface hydrophilic and resistant to protein adsorption [19,69–73].

Strategies employed in attempting to solve PDMS hydrophobicity include surface activation methods such as: oxygen plasma; UV/ozone treatments; corona discharges, which are widely used for PDMS surface oxidation to promote microchannel wettability. The main benefits of these methods are the short treatment time and easy operation; however, the PDMS surface recovers its hydrophobicity when in contact with air within a few minutes [74–76]. The hydrophilic treatments and some examples are discussed further in Section 5.

Another method is physisorption, which is a simple and efficient approach that relies on surface hydrophobic or electrostatic interactions. This method includes the following techniques: layer-by-layer deposition; non-ionic surfactants; charged polymers. The disadvantages are the lack of covalent bonds between PDMS and surface modifiers, which lead to the loss of modifiers quickly through desorption [77–79].

In order to improve the difficulties encountered in physisorption, chemical modification methods allow for maintaining a long-term stability of the modified surface. These methods include: chemical vapor deposition, surface segregation and self-assembled monolayers, silanization, and polymer brushes via grafting methods [1,62,80–82].

Adding waxes such as paraffin or beeswax to PDMS has been demonstrated to potentially increase the corrosion resistance, hydrophobicity and thermal and optical properties of PDMS, which is useful in applications such as sensors, wearable devices and superhydrophobic coating [83].

Although the methods listed above have been successful in improving the hydrophilicity of the PDMS surface, they have some limitations, such as chemical instability, the need for specific equipment, limited manufacturing process for large scale and some methods cause loss of transparency, loss of mechanical properties and do not provide the hydrophilic surface for a long period of time [62]. Considering these facts, the work of Gökaltun et al. [84] presents a simplified method of easy manufacture, which uses copolymers composed of poly(ethylene glycol) and PDMS segments (PDMS-PEG) to reduce the hydrophobicity of PDMS without changing its transparency, biocompatibility and mechanical properties, with a durability of 20 months.

3. PDMS Manufacturing Process

Sylgard® 184 Silicone Elastomer Kit is the most used commercial PDMS. It consists of a monomer and a curing agent, which are usually combined at a weight ratio of 10:1. The compound is mixed and then degassed with a desiccator in order to prevent the formation of micro-bubbles. The PDMS solution is poured over the master mould and then cured in the oven [23]. The curing time depends on the temperature of the oven and on the size of the PDMS sample. The higher the hardening temperature, the less time it will take for the PDMS to cure. After the curing process, the piece is taken out of the mould [57]. Note that, for very specific applications and complex geometries, it is usually advised to perform the curing process at room temperature for at least 48 h [55,85]. In Table 2 are listed curing times and temperatures recommended by the manufacturer.

Table 2. Recommended curing times and temperatures to produce PDMS samples [57].

Temperature (°C)	Time
25	48 h
100	35 min
125	20 min
150	10 min

The monomer and the curing agent can be mixed at a different ratio besides the 10:1 [86] and, as a consequence, some properties change, namely, mechanical [87], optical [88] and gas permeability [89]. Mixing at a higher ratio of cure agent results in a faster hardening time, in a less sticky cured PDMS and in a more fragile PDMS sample. In contrast, mixing with less cure agent results in a longer hardening time, in a stickier cured PDMS and in better mechanical properties. Khanafer et al. [87] found that elastic modulus increases as the mixing ratios increase up to 9:1, after which the elastic modulus starts to decrease as the mixing ratio continues to increase.

4. Methods to Characterize PDMS

A wide range of tests are performed to characterize elastomers. Some common tests are scanning electron microscopy, gravimetry, goniometry, nanoindentation, tensile test, X-ray photoelectron spectroscopy and Fourier Transform infrared spectroscopy [21]:

- Scanning electron microscopy (SEM) allows thickness measurement and qualitative characterization of PDMS samples [18,42,90–92];
- Gravimetry is a method based on gravitational techniques to quantify changes in PDMS sample weight. For example, this method is useful when it is needed to verify if there was or not degradation of the PDMS after chemical immersion [93];
- In order to obtain information on surface hydrophilicity, a goniometry test is performed. Micro water droplets are dropped on the PDMS surface and then the contact angle is measured. This technique allows for verification of if there was or not a change in the wettability of the PDMS after certain treatments [19,39,42,94];
- Nanoindentation offers the possibility of studying mechanical properties of the outermost layer of PDMS, which is susceptible to destruction due to different treatments, such as UV irradiation [95];
- Tensile testing allows Young Modulus measurement of PDMS. The Young Modulus can be affected by treatments that may be applied to PDMS, by hardening temperature and time, and by the mixing ratio used to fabricate the PDMS samples [42,96,97];
- X-ray photoelectron spectroscopy (XPS) is a technique based on the photoelectric effect, which allows identification of the elemental composition of the material. This method is useful when it is needed to verify if any changes in surface composition occurred after the PDMS received any treatment [38,39,98];
- Fourier Transform infrared spectroscopy (FTIR) is a method used to obtain the infrared spectrum of absorption or transmission of the PDMS sample. This technique allows examination of the effect of some treatment on the cross-linking of PDMS [38,42,99].

5. PDMS Microfabrication

PDMS is patterned through commonly used microfabrication techniques, such as soft-lithography and spin coating. However, especially due to its hydrophobic nature, some of the techniques must be employed alongside with hydrophilic treatments, such as oxygen plasma. Soft-lithography, which is a group of techniques that use patterned elastomers as stamp, mould or mask to generate micropatterns, was developed to allow processing elastomers [100]. However, the fabrication of the most microfluidic devices still relies on photolithography for fabricating SU-8 masters that usually serve as the PDMS mould [20]. Photolithography is a microfabrication technique used to process photoresists, commonly employed in CMOS microelectronics fabrication [101]. The soft-lithography can be performed in several types, such as microcontact printing (μCP), replica moulding (REM), micro-transfer moulding (μTM), micro-moulding in capillaries (MIMIC), solvent-assisted micro-moulding (SAMIM), phase-shift photolithography, cast moulding, embossing and injection. Some of these techniques are briefly described below [100]:

1. Microcontact printing: uses the relief pattern on the surface of a PDMS stamp to form patterns of self-assembled monolayers (SAMs) on the surfaces of substrates by contact;
2. Replica moulding: replicates the relief pattern on the surface of a PDMS mould by using this structure as a mould for forming structures in a second UV-curable (or thermally curable) prepolymer;
3. Micro-transfer moulding: a thin layer of liquid prepolymer is applied to the patterned surface of a PDMS mould. It is then placed in contact with the surface of a substrate and the liquid prepolymer is cured to a solid. After peeling off the mould, a patterned micro-structure is left on the surface of the substrate;
4. Micro-moulding in capillaries: a PDMS mould is placed on the surface of a substrate to form a network of empty channels between them. The channels are filled with a

low viscosity prepolymer, which is then cured to a solid. The mould is removed and a patterned micro-structure is left on the surface of the substrate;

5. Solvent-assisted micro-moulding: a PDMS mould is wetted with a solvent, and it is placed in contact with a substrate (typically an organic polymer). The solvent starts to dissolve the substrate into a fluid or gel that is moulded against the relief structures in the mould. When the fluid solidifies, it forms a pattern relief structure complementary to that in the surface of the mould.

The soft-lithography process begins with the preparation of the elastomeric stamp or the mould by cast moulding. Most of the time, cast moulding implies the use of photolithographic techniques to fabricate the master. PDMS is the most widely used elastomer for this process because of its outstanding properties: low interfacial free energy, it does not swell with humidity, good thermal stability, optical transparency, isotropy and homogeneity [100].

Additionally, spin coating is a common microfabrication method for producing polymer films of controlled and uniform thickness. In this process, a liquid film is spread by centrifugal force onto a rotating substrate. This technique is commonly used for deposition of polymer resist layers in the photolithographic processing of a master mould. It is formulation dependent: increased amounts of cross-linker agent in the formulation decrease film thickness [21,102].

The hydrophobic nature of PDMS brings, in some cases, limitations in the microfabrication processes. There are applications, such as cell culture, immunoassay and biomolecule separation, where the modification of the hydrophobic surface of native PDMS to a hydrophilic surface is indispensable. For example, when endothelial cell seeding is needed, hydrophilic modification of the PDMS surface is indispensable for a successful seeding [103]. However, it is important that the hydrophilic treatment does not affect its transparency, as transparency is a key property that makes PDMS the material of choice for certain applications. Oxygen plasma is the most employed treatment that leads to an increase in PDMS surface hydrophilicity because of its short treatment time, its easy operation and that it does not affect the PDMS transparency [21,65,104,105]. However, this treatment is also known for losing its effects within minutes after exposure to air. For this reason, a variety of well-studied treatments have emerged for this purpose [19,69–73]. Additionally, some articles reported that oxygen plasma may damage PDMS surface [62,106] and, therefore, Shin et al. reported three different treatments that do not require oxygen plasma pre-treatment, including Teflon coating, commercially available water-repellents and perfluorodecyltrichlorosilane (FDTS) [107]. The authors showed that the Teflon and the water-repellent decreased the hydrophobicity of PDMS with great chemical stability and without significantly affecting its transparency. UV/ozone treatments and corona discharge are also commonly employed hydrophilic treatments; however, as with oxygen plasma treatment, PDMS quickly recovers its hydrophobicity [62]. There have been efforts to improve some of these treatments; however, the best way to achieve an effective and long-lasting treatment seems to be the combination of a surface activation with a covalent surface functionalization [19,106]. For example, Zhao et al. [108] proposed a method where PDMS is firstly activated by oxygen plasma treatment and then it is coated with a zwitterionic poly(methacrylate) copolymer (PMGT). This method allowed for decreasing the water contact angle (WCA) of native PDMS from 108° to 30°, with a duration of at least 200 h. Further, Zhou et al. [19] suggest a combination of gas-phase with wet chemical methods in order to achieve a better surface stability in a shorter treatment time. Examples of these treatments are the combination of UV or plasma treatment and silanization, the combination of UV or plasma treatment and graft polymerization and the combination of plasma treatment and layer-by-layer (LBL) assembly.

Sterilization is a required procedure for most biomedical applications. In some cases, this process must be done alongside the microfabrication process. There are three mainly used sterilization methods: cleaning with ethanol, ultraviolet light exposure and the steam autoclave procedure. Sterilization does not significantly affect PDMS hydrophobicity. However, steam autoclaving increases the storage modulus and ultimate tensile stress [21,109–114].

6. PDMS Applications

6.1. PDMS-Based Microchip

Microfluidic devices have been widely studied and developed and, in order to take them to the market, they must be low-cost and capable of mass production. The use of PDMS to fabricate these devices makes it possible to achieve those goals. Nowadays, there are a variety of PDMS-based microchips that have been developed, most of them alongside glass [115–121]. The combination of PDMS and glass has been employed with great results. In Figure 1 is presented a schematic illustration of the fabrication of a glass/PDMS microchip. For example, Schöning et al. [116] developed a PDMS/glass separation microchip, based on typical semiconductor-compatible production methods, and which provides a simplification of the electrophoresis-based biosensor set-up.

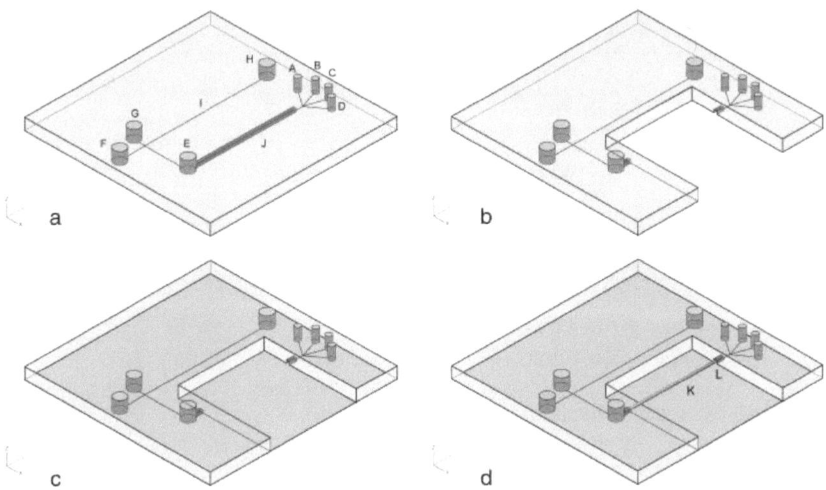

Figure 1. Schematic illustration of the fabrication for glass/PDMS microchip: (**a**) PDMS layer fabricated by replica moulding; (**b**) part of SPE channel in PDMS layer was cut off; (**c**) PDMS layer was sealed with the thin glass cover slip; (**d**) MISPE monolithic capillary column was coupled with glass/PDMS chip to form the final chip. A, B, C, D: holes, E: sample reservoir, F: buffer reservoir, G: sample waste reservoir, H: buffer waste reservoir, I: separation channel, J: SPE channel, K: MISPE monolithic capillary column and L: epoxy glue. Reprinted with permission from reference [115]. Copyright 2020 Elsevier.

Temperature gradient generation is a commonly used process in microfluidics. Ha et al. [122] presented a PDMS microchip that allows temperature gradient generation using sound waves as a heating mechanism. The use of PDMS allowed the fabrication of a transparent, dynamic, inexpensive and easy-to-fabricate system.

The hydrophobic nature of PDMS usually brings issues to the microchips. Qiang Niu et al. [120] developed a PDMS/glass microchip for PCR; however, the team came across the formation of bubbles on the PDMS surface during the sample loading. To overcome this, they implemented an irreversible bonding and sealing between the glass and PDMS. Additionally, protein adsorption occurred on the chip surface, which was overcome by treating the surface with BSA (Bovine Serum Albumin). Table 3 presents a list of PDMS-based microchips, as well as the motivations that led the authors to use PDMS in their devices.

Table 3. Applications of PDMS-based microchips and respective motivations for using PDMS.

Application	PDMS Preparation	Motivations for Using PDMS	Reference
On-line sample pre-treatment and contactless conductivity detection	Mixing ratio—10:1, *w/w* Degassing time—20 min Curing temperature—80 °C Curing time—30 min Oxygen plasma treatment for 1–2 min	Low-cost, easy manufacture, suitability for mass production, transparency and elasticity.	[115]
Genetic analysis by functional integration of polymerase chain reaction (PCR) and capillary gel electrophoresis (CGE)	Mixing ratio—10:1, *w/w* Degassing time—15 min Curing temperature—65 °C Curing time—1 h Post-curing temperature—135 °C Post-curing time—15 min Hydrophilic treatment with HCl solution at 25 °C for 4 h	Low-cost, suitability for microscale moulding, high reproducibility on a micrometre scale, high gas permeability, low thermal conductivity and transparency.	[118]
Polymerase chain reaction (PCR)	Mixing ratio—10:1, *w/w* Curing temperature—95 °C Curing time—30 min	Low thermal conductivity, simple fabrication, low-cost, disposability, biocompatibility, irreversible bonding with glass and transparency.	[120]
Electrophoresis device for continuous on-line in vivo monitoring of micro dialysis samples	Mixing ratio—10.5:1.5, *w/w* 5 mm-thick layer curing temperature—90 °C 5 mm-thick layer curing time—25–30 min 1 mm-thick layer curing temperature–90 °C 1 mm-thick layer curing time—15–18 min Post-curing temperature—85 °C Post-curing time—overnight	Easy manufacture, good reproductivity and transparency.	[121]
Generation of temperature gradient	Mixing ratio—10:1, *w/w*	Low-cost, transparency, easy manufacture and low thermal conductivity	[122]

6.2. PDMS Biomodels for Hemodynamic Studies

As mentioned above, PDMS can be very useful in the fabrication of microchips that allow analysing samples. Reports on the use of this material for the replication of cardiovascular flow are also found in the literature. This type of application allows a better understanding and study of cardiovascular diseases, such as aneurysms.

An aneurysm is characterized by artery wall weakness, which can lead to artery rupture and, consequently, to death. Hemodynamic studies have been done to understand aneurysms; however, they cannot explain the mechanical effects on the expansion of the aneurysm walls [49]. To understand these mechanical effects, studies were conducted where an intercranial aneurysm model was developed using PDMS to simulate the mechanical behaviour of blood vessels [49,123]. PDMS is a well-suited material for this purpose due to its hyperelastic behaviour, which is very similar to that of blood vessels, and the ability to make circular microchannels. Another advantage of using PDMS is that it is transparent, which facilitates monitoring of the blood flow. Additionally, recent studies show that is possible to seed a culture of endothelial cells on the microchannels' walls, which allows the creation of a very similar environment to that found in microcirculation [90,124,125]. Lima et al. [126] proposed a microfluidic device containing rectangular microchannels in PDMS, where in vitro blood flow measurements were conducted by means of a confocal micro-PIV system. The authors demonstrated that, by using soft-lithography, it is possible to produce precise and reproductible rectangular microchannels and to perform detailed blood flow studies. The same authors have performed a similar study, this time by using circular PDMS microchannels [127]. Although there are already several studies using circular microchannels [127–130], the majority of the PDMS microchannels used to study in vitro blood flow phenomena have rectangular cross sections. Hence, by using rectangular PDMS microchannels, several research works have been performed on different kinds of constrictions to study the deformability behaviour of blood cells [131–136] and air bubbles [137,138]. Cell deformability is a biomarker which can be used to distinguish between healthy and diseased cells. Microfluidic models have been developed in order to better understand and, consequently, diagnose diseases such as malaria [139,140], cancer [141,142] and end-stage kidney disease [143]. Most studies aim to better understand red blood cells (RBC); however, Rodrigues et al. [144] developed a novel integrative microfluidic device which is capable of assessing the deformation index of both white blood cells (WBC) and RBC. The same author also presented a microfluidic tool to study the hemocompatibility of nanoparticles synthesized for theragnostic applications [145]. Additionally, by using microchannels having bifurcations and confluences, several studies have been carried out to better understand the influence of these complex geometries on blood flow behaviour [146–152].

Rectangular microchannels are the most common geometry obtained by soft-lithography. However, this kind of geometry can lead to some erratic measurements because the shear stress imposed on the cell is different and, consequently, the pressure build-up in the channel is not the same as if it were built-up in a circular section [124]. Hence, studies have been conducted to establish methods that allow the construction of circular microchannels of PDMS. For example, Fiddes et al. [124] proposed a method which begins by fabricating rectangular microchannels using soft-lithography techniques, followed by the introduction of a stream gas and a solution of the silicone oligomer in an organic solvent. Then, through the polymerization of the oligomer and the removal of the solvent, the authors demonstrated the ability to control the shape and the diameter of the microchannel's cross-section. Additionally, Choi et al. [90] showed that, combining soft-lithography techniques with the reflow phenomenon of a positive photoresist, it is possible to generate circular PDMS microfluidic channels. In Figure 2 are presented some examples of biomodels for hemodynamic studies.

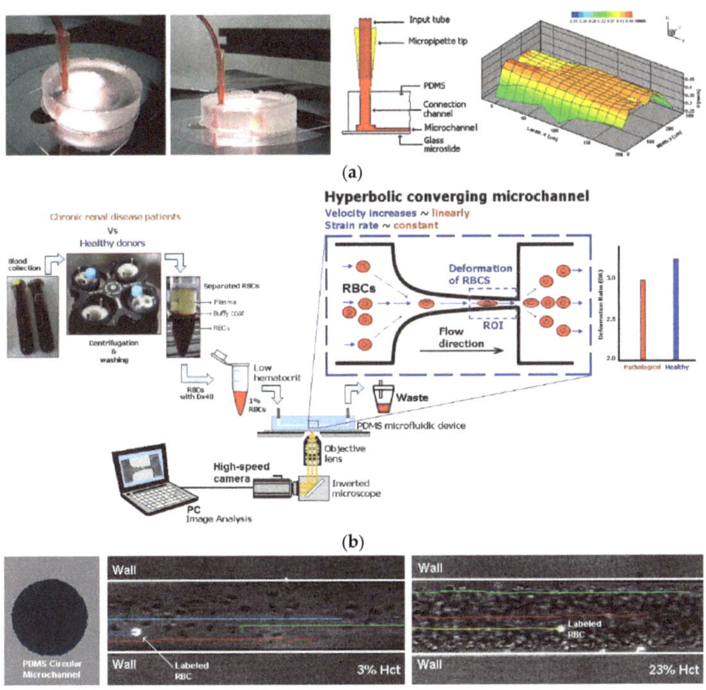

Figure 2. Example of PDMS biomodels for hemodynamic studies: (**a**) rectangular PDMS microchannel to study in vitro blood and ensemble velocity profiles (U) obtained in the middle plane by means of a confocal micro-PIV system (adapted from [126]); (**b**) schematic diagram of the blood collection and cells deformability tests in PDMS microfluidic device (from [143]); (**c**) circular PDMS microchannels to study in vitro blood behavior (adapted from [152]).

It is important to notice that, despite the ability to mimic the cardiovascular vessels behaviour through PDMS microchannels, there would always be missing points. For that reason, it is of great importance to combine the PDMS micro devices with well-suited measurement techniques. Rodrigues et al. [123] proposed the use of the Digital Image Correlation (DIC) method, which proved to be suitable to study small displacements happening in in vitro models. A summary of the advantages and limitations of some of the methods used to fabricate microchannels is presented in Table 4.

Table 4. Advantages and limitations of techniques used to fabricate microchannels.

Geometry	Method	Advantages	Limitations	Application	Reference
Rectangular	Soft lithography	Generation of precise, reproducible and versatile microchannels; Precise control of experimental parameters and accurate measurements; Inexpensive, simple and rapid method.	Different geometry from in vivo microvessels; Difficulties in achieving stable cell seeding at the corners of the channel.	Integration of confocal micro-PIV with a PDMS microchannel to obtain blood velocity profiles	[126]
Circular	Wire casting technique	Simple and inexpensive method; Possibility of fabricating microchannels with different diameters; No need for a clean room or specialized equipment.	It is not possible to generate well-defined complex structures, such as bifurcations.	In vitro hemodynamic studies	[127]
	Partially cured PDMS combined with thermal air expansion molding	Inexpensive and simple method; Possibility of fabricating multiple diameters of circular channel from 100 µm to 500 µm and different cross-sections.	It can be hard to fabricate a perfect circular channel.	Evaluate the clotting events in pathological vessels and testing device for antiplatelet and anticoagulant therapeutics	[128]
	Combination of soft lithography with the reflow phenomenon of a positive photoresist	Simple and efficient method; Possibility of fabricating microchannels with multiple diameters (from 100 µm to 400 µm) and various channel designs.	It can be hard to control the thickness of the photoresist, leading to a difficulty in generate perfect circular channels; Bonding the two semi-circular channels perfectly can be challenging.	This method allows endothelial cells culture, making this project suitable for drug screening and chemical/biological diagnostics	[90]
	Reshaping rectangular microchannels through polymerization of the liquid silicone oligomer around a gas steam	Ability of controlling the diameter from 40 µm to 100 µm; Possibility of fabricating constrictions.	Relatively complex and expensive method; Difficulty in controlling the exact diameter of the channel.	Mimic in vivo systems for cell flow studies	[124]

6.3. PDMS-Based Blood Analogues

Blood analogues are fluids commonly used to perform hemodynamic experiments due mainly to safety problems related to the use of real blood in these experiments. Initially, blood analogues were simple fluids composed by mixtures of glycerol and water or by xanthan gum diluted in glycerine and/or water [153,154]. However, by using these kinds of blood analogues, it is not possible to study different kinds of flow phenomena that happen at the micro scale level, such as the cell-free layer, plasma skimming and cell margination [101,155,156]. These microcirculation phenomena do not happen by using blood analogue fluids without solid elements, such as microparticles and microcapsules. Hence, during the past years, several works have been developing different kinds of particulate blood analogue fluids containing microparticles with varying stiffness, shape and size for biomedical applications [157–166].

Due to its unique mechanical properties, PDMS has also been used to produce flexible microparticles to be used in blood analogue fluids. Recently, Muñoz-Sánchez et al. [167] proposed a flow-focusing technique to produce flexible PDMS microparticles for biomedical applications (Figure 3). The PDMS microparticles were produced by using different kinds of ratios (base/curing agent), and rheological measurements performed with a ratio of 6:4 have demonstrated the ability to reproduce the steady shear viscosity curve of ovine RBCs suspended in Dextran 40 [167–169]. Although it is possible to produce flexible PDMS microparticles with a high degree of monodispersity by using the flow-focusing technique, the production rate is relatively low. In order to overcome this limitation, Choi et al. [170] and Lopez et al. [171] have proposed a simple emulsification technique to obtain a mass production of PDMS microparticles. More recently, Carneiro et al. [172] have developed another method, based on a multi-stage membrane emulsification process, to obtain high throughput production of PDMS microparticles. The development of blood analogue fluids with PDMS microparticles that mimic the behaviour of RBCs is still at an early stage of development. The most critical challenges that need to be solved are the mass production of monodisperse PDMS microparticles, stiffness, aggregation and fast agglomeration of the PDMS particles within microchannels with complex geometries, such as constrictions and bifurcations.

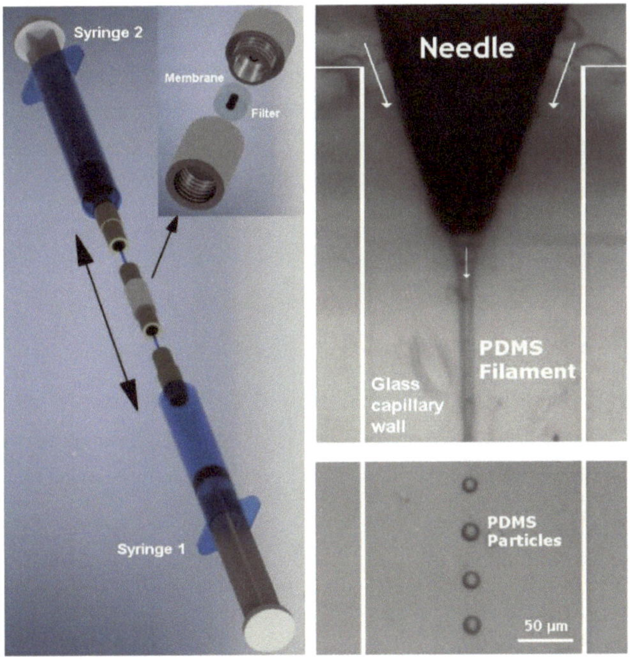

(a) (b)

Figure 3. Flexible PDMS microparticles to be used in blood analogue fluids for biomedical applications, produced by (**a**) a two-syringe membrane emulsification technique (adapted from [171]); (**b**) a flow-focusing technique (adapted from [173]).

6.4. PDMS-Based Coatings for Medical Implants

PDMS has been widely studied to integrate medical implants, especially due to its biocompatibility. Such implants are usually fabricated with biomedical grade metals (e.g., tantalum, zirconium, niobium), as well as titanium and its alloys [174]. However, these present some limitations concerning blood compatibility, bone conductivity and bioactivity [38]. When developing an implant, some important aspects should be taken in consideration: biocompatibility, osseointegration, corrosion resistance and micro-invasiveness. Osseointegration is related to the effective linkage between the metal and the bone. A weak bonding can lead to the formation of biofilms on implants, which can cause infections. Recent studies have demonstrated that the surface modification of implants, in order to achieve nano-/microscale features, brings great advantages concerning osseointegration [39].

The creation of microscale features on ceramics or polymers is simpler than on metal. Considering the fact that PDMS allows the fabrication of hydrophobic and smooth surfaces has led to their use for developing coatings that help in the osseointegration of implants [38–42]. Rossi de Aguiar et al. [38] studied a sol-gel coating based on PDMS for metallic surfaces such as titanium and stainless steel. The authors demonstrated that the hydrophobic nature of PDMS allows the formation of an anti-biofouling surface, preventing the bacterial adhesion. Additionally, Tran et al. [39] developed a coating that involved the hydrolysis and co-condensation of PDMS and tantalum (Ta) ethoxide to produce tantalum oxide. This PDMS hybrid material has biocompatibility and corrosion resistance properties, which allowed a great osseointegration. The integration of nanoparticles, such as CuO, has been proven to improve the antibacterial characteristic of PDMS-based coatings, as demonstrated by Tavakoli et al. [42]. Table 5 comprises a list of some PDMS-based coatings that have been developed in the past years.

Table 5. Developments and applications of PDMS-based coatings.

Application	PDMS Preparation	Motivation for Using PDMS	Reference
Urethanes PDMS-based hybrid coating for metallic dental implants	Hybrid urethanesil (PDMSUr) synthesized by ring opening polymerization of a bis(cyclic carbonate) derived from PDMS. Curing temperature—60 °C Curing time—24 h	Create hydrophobic and smooth surfaces, with less adhesion of bacteria, capable of adhering to tissue cells such as fibroblasts and osteoblasts.	[38]
Tantalum oxide-PDMS hybrid coating for medical implants	Modified sol-gel synthesis method, Tantalum oxide-PDMS solutions (10%, *v*/*v*). Curing temperature—room temperature Curing time—15 min	Medical grade PDMS has functional groups to bind to reactive surfaces such as activated metals or polymers. Ability to create micrometer-thick coatings.	[39]
Bioactive $CaO\text{-}SiO_2$-PDMS coatings	Sol-gel dip-coating method. The produced coatings were kept at room temperature for 24 h for gelation. Curing temperature—150 °C Curing time—24 h	Mechanical properties and elasticity of PDMS	[40]
PDMS-based coating for a bladder volume monitoring sensor	Mixing ratio—10:2 (*w*/*w*) Curing temperature—80 °C Curing time—2 h	Biocompatibility, 10:2 ratio to increase tensile strength and improve Young's modulus	[41]
CuO-PDMS-SiO_2 coatings	Mixing ratio—10:1 (*w*/*w*) Curing temperature—150 °C Curing time—90 min	Improved biocompatibility, corrosion resistance and antibacterial property	[42]

7. Conclusions and Further Perspectives

Microchips for biomedical applications are devices that allow monitoring and analysis of samples. The use of PDMS in these devices offers great advantages such as optical transparency, being easy-to-manufacture and having a low-cost, which are important requirements when fabricating microchips. Additionally, the permeability to gases is a unique advantage to culture living cells in closed microchannels, a task that is extremely complex to achieve in glass microchannels. However, the hydrophobic nature of PDMS brings some limitations during the fabrication and flow transport phenomena, especially for biological applications. Developments of treatments that contradict the hydrophobic property have been made and these limitations are easily overcome by applying simple and fast hydrophilic treatments to PDMS. The lack of industrial processes to manufacture PDMS is still an issue. There are already methods that allow good replications of microfabricated PDMS; however, they are far from an industrial scale.

Replication of the cardiovascular system using PDMS microchannels is on a good path to be an application well-suited for the study of cardiovascular diseases. The hyper-elastic behaviour and transparency are great advantages that make PDMS the chosen material in these types of applications. Herein, the hydrophobic nature of PDMS can be a limitation as well, in the blood flow itself but also when it is intended to grow endothelial cell cultures on its walls.

Additionally, PDMS plays an important role in medical implant applications, especially due to its biocompatibility and hydrophobic nature. These characteristics allow the production of antibacterial coatings for implants, which is a requirement when developing implants. PDMS also allows the production of smooth surfaces through processes of microfabrication that help in the osseointegration of the implant in the body. Although PDMS coatings are already available on the market, there are more developments that can be made to increase their features and durability.

It is interesting to note that the hydrophobic nature of PDMS can be a limitation in some applications, such as microchips and microchannel fabrication, but a great advantage in others, such as in implant coatings, solar panels and face masks.

In summary, PDMS opens a wide range of possibilities to make great developments in biomedical applications. With regards to further work, it is important to continue studying more methods to produce PDMS-based devices on a larger scale which would further enable these devices to reach the market. Additionally, the currently available PDMS hydrophilic treatments need further developments and improvements as, most of the time, they do not last long. Hence, it is important to develop new methods or improve the existing ones in order to achieve a higher permanent hydrophilic feature for PDMS.

Author Contributions: Conceptualization, I.M., A.S. and P.S.; writing—original draft preparation, I.M., A.S., P.S. and J.R.; writing—review and editing, G.M., E.M.S.C., J.R. and R.L.; supervision, G.M. and R.L.; funding acquisition, G.M., J.R. and R.L. All authors have read and agreed to the published version of the manuscript.

Funding: This work was supported by projects NORTE-01-0145-FEDER-029394 and NORTE-01-0145-FEDER-030171, through Programa Operacional Regional do Norte–Norte Portugal Regional Operational Programme (NORTE 2020), under the PORTUGAL 2020 Partnership Agreement through the European Regional Development Fund (FEDER) and by Fundação para a Ciência e Tecnologia (FCT), IP, project reference PTDC/EMD-EMD/29394/2017 and PTDC/EME-SIS/30171/2017.

Institutional Review Board Statement: Not applicable.

Informed Consent Statement: Not applicable.

Data Availability Statement: Not applicable.

Acknowledgments: The authors also acknowledge the strategic grants by the projects UIDB/04436/2020, UIDP/04436/2020, UIDB/00690/2020, UIDB/04650/2020 and UIDB/04077/2020 from FCT.

Conflicts of Interest: The authors declare no conflict of interest.

References

1. Poll, M.; Zhou, F.; Ramstedt, M.; Hu, L.; Huck, W. A Self-assembly approach to chemical micropatterning of Poly(dimethylsiloxane). *Angew. Chem. Int. Ed.* **2007**, *46*, 6634–6637. [CrossRef] [PubMed]
2. Berthier, E.; Young, E.W.K.; Beebe, D. Engineers are from PDMS-land, Biologists are from Polystyrenia. *Lab Chip* **2012**, *12*, 1224–1237. [CrossRef]
3. Merkel, T.C.; Bondar, V.I.; Nagai, K.; Freeman, B.D.; Pinnau, I. Gas sorption, diffusion, and permeation in Poly(dimethylsiloxane). *J. Polym. Sci. Part B Polym. Phys.* **2000**, *38*, 415–434. [CrossRef]
4. Kuddannaya, S.; Bao, J.; Zhang, Y. Enhanced In Vitro biocompatibility of chemically modified Poly(dimethylsiloxane) surfaces for stable adhesion and long-term investigation of brain cerebral cortex cells. *ACS Appl. Mater. Interfaces* **2015**, *7*, 25529–25538. [CrossRef]
5. Lee, S.; Shin, H.-J.; Yoon, S.-M.; Yi, D.K.; Choi, J.-Y.; Paik, U. Refractive index engineering of transparent ZrO_2–polydimethylsiloxane nanocomposites. *J. Mater. Chem.* **2008**, *18*, 1751–1755. [CrossRef]
6. Johnston, I.D.; Tracey, M.C.; Davis, J.B.; Tan, C.K.L. Micro throttle pump employing displacement amplification in an elastomeric substrate. *J. Micromech. Microeng.* **2005**, *15*, 1831–1839. [CrossRef]

7. Dardouri, M.; Bettencourt, A.; Martin, V.; Carvalho, F.A.; Santos, C.; Monge, N.; Santos, N.C.; Fernandes, M.H.; Gomes, P.S.; Ribeiro, I.A.C. Using plasma-mediated covalent functionalization of rhamnolipids on polydimethylsiloxane towards the antimicrobial improvement of catheter surfaces. *Mater. Sci. Eng. C* **2021**, 112563. [CrossRef]
8. Kumar, R.; Sahani, A. Role of superhydrophobic coatings in biomedical applications. *Mater. Today* **2021**, *45*, 5655–5659. [CrossRef]
9. Wu, X.; Kim, S.-H.; Ji, C.-H.; Allen, M. A solid hydraulically amplified piezoelectric microvalve. *J. Micromech. Microeng.* **2011**, *21*, 95003–95011. [CrossRef]
10. Bozukova, D.; Pagnoulle, C.; Jérôme, R.; Jérôme, C. Polymers in modern ophthalmic implants—Historical background and recent advances. *Mater. Sci. Eng. R Rep.* **2010**, *69*, 63–83. [CrossRef]
11. Yu, H.; Zhou, G.; Sinha, S.K.; Chau, F.S.; Wang, S. Lens integrated with self-aligned variable aperture using pneumatic actuation method. *Sens. Actuators A Phys.* **2010**, *159*, 105–110. [CrossRef]
12. Doutel, E.; Viriato, N.; Carneiro, J.; Campos, J.B.L.M.; Miranda, J.M. Geometrical effects in the hemodynamics of stenotic and non-stenotic left coronary arteries-numerical and in vitro approaches. *Int. J. Numer. Methods Biomed. Eng.* **2019**, *35*, e3207. [CrossRef] [PubMed]
13. Usmani, A.; Muralidhar, K. Flow in an intracranial aneurysm model: Effect of parent artery orientation. *J. Vis.* **2018**, *21*, 795–818. [CrossRef]
14. Kim, S.-J.; Lee, D.-S.; Kim, I.-G.; Sohn, D.-W.; Park, J.-Y.; Choi, B.-K.; Kim, S.-W. Evaluation of the biocompatibility of a coating material for an implantable bladder volume sensor. *Kaohsiung J. Med. Sci.* **2012**, *28*, 123–129. [CrossRef]
15. Carta, R.; Jourand, P.; Hermans, B.; Thoné, J.; Brosteaux, D.; Vervust, T.; Bossuyt, F.; Axisa, F.; Vanfleteren, J.; Puers, R. Design and implementation of advanced systems in a flexible-stretchable technology for biomedical applications. *Sens. Actuators A Phys.* **2009**, *156*, 79–87. [CrossRef]
16. Fujii, T. PDMS-based microfluidic devices for biomedical applications. *Microelectron. Eng.* **2002**, *61–62*, 907–914. [CrossRef]
17. Raj, M.K.; Chakraborty, S. PDMS microfluidics: A mini review. *J. Appl. Polym. Sci.* **2020**, *137*, 48958. [CrossRef]
18. Chen, W.; Lam, R.H.W.; Fu, J. Photolithographic surface micromachining of polydimethylsiloxane (PDMS). *Lab Chip* **2012**, *12*, 391–395. [CrossRef]
19. Zhou, J.; Ellis, A.V.; Voelcker, N.H. Recent developments in PDMS surface modification for microfluidic devices. *Electrophoresis* **2010**, *31*, 2–16. [CrossRef]
20. Weibel, D.B.; DiLuzio, W.R.; Whitesides, G.M. Microfabrication meets microbiology. *Nat. Rev. Microbiol.* **2007**, *5*, 209–218. [CrossRef]
21. Mata, A.; Fleischman, A.J.; Roy, S. Characterization of Polydimethylsiloxane (PDMS) Properties for Biomedical Micro/Nanosystems. *Biomed. Microdevices* **2005**, *7*, 281–293. [CrossRef]
22. Ashraf, M.W.; Tayyaba, S.; Afzulpurkar, N. Micro Electromechanical Systems (MEMS) based microfluidic devices for biomedical applications. *Int. J. Mol. Sci.* **2011**, *12*, 3648–3704. [CrossRef]
23. Schneider, F.; Fellner, T.; Wilde, J.; Wallrabe, U. Mechanical properties of silicones for MEMS. *J. Micromech. Microeng.* **2008**, *18*, 065008. [CrossRef]
24. Bubendorfer, A.; Liu, X.; Ellis, A.V. Microfabrication of PDMS microchannels using SU-8/PMMA moldings and their sealing to polystyrene substrates. *Smart Mater. Struct.* **2007**, *16*, 367–371. [CrossRef]
25. Pinto, V.C.; Sousa, P.J.; Cardoso, V.F.; Minas, G. Optimized SU-8 Processing for low-cost microstructures fabrication without cleanroom facilities. *Micromachines* **2014**, *5*, 738–755. [CrossRef]
26. Shakeri, A.; Khan, S.; Didar, T.F. Conventional and emerging strategies for the fabrication and functionalization of PDMS-based microfluidic devices. *Lab Chip* **2021**, *21*, 3053–3075. [CrossRef]
27. Jo, M.C.; Guldiken, R. Effects of polydimethylsiloxane (PDMS) microchannels on surface acoustic wave-based microfluidic devices. *Microelectron. Eng.* **2014**, *113*, 98–104. [CrossRef]
28. Levitt, M.R.; Mandrycky, C.; Abel, A.; Kelly, C.M.; Levy, S.; Chivukula, V.K.; Zheng, Y.; Aliseda, A.; Kim, L.J. Genetic correlates of wall shear stress in a patient-specific 3D-printed cerebral aneurysm model. *J. Neurointerv. Surg.* **2019**, *11*, 999–1003. [CrossRef] [PubMed]
29. Doutel, E.; Carneiro, J.; Oliveira, M.; Campos, J.B.L.M.; Miranda, J. Fabrication of 3d mili-scale channels for hemodynamic studies. *J. Mech. Med. Biol.* **2014**, *5*, 21. [CrossRef]
30. Doutel, E.; Carneiro, J.; Campos, J.B.L.M.; Miranda, J.M. Experimental and numerical methodology to analyze flows in a coronary bifurcation. *Eur. J. Mech. B Fluids* **2018**, *67*, 341–356. [CrossRef]
31. Geoghegan, P.H.; Buchmann, N.A.; Spence, C.J.T.; Moore, S.; Jermy, M. Fabrication of rigid and flexible refractive-index-matched flow phantoms for flow visualisation and optical flow measurements. *Exp. Fluids* **2012**, *52*, 1331–1347. [CrossRef]
32. Ford, M.D.; Nikolov, H.N.; Milner, J.S.; Lownie, S.P.; Demont, E.M.; Kalata, W.; Loth, F.; Holdsworth, D.W.; Steinman, D.A. PIV-measured versus CFD-predicted flow dynamics in anatomically realistic cerebral aneurysm models. *J. Biomech. Eng.* **2008**, *130*, 21015. [CrossRef]
33. Brindise, M.C.; Rothenberger, S.; Dickerhoff, B.; Schnell, S.; Markl, M.; Saloner, D.; Rayz, V.L.; Vlachos, P.P. Multi-modality cerebral aneurysm haemodynamic analysis: In vivo 4D flow MRI, in vitro volumetric particle velocimetry and in silico computational fluid dynamics. *J. R. Soc. Interface* **2019**, *16*, 20190465. [CrossRef] [PubMed]
34. Amili, O.; Golzarian, J.; Coletti, F. In Vitro Study of particle transport in successively bifurcating vessels. *Ann. Biomed. Eng.* **2019**, *47*, 2271–2283. [CrossRef]

35. Li, Y.; Verrelli, D.I.; Yang, W.; Qian, Y.; Chong, W. A pilot validation of CFD model results against PIV observations of haemodynamics in intracranial aneurysms treated with flow-diverting stents. *J. Biomech.* **2020**, *100*, 109590. [CrossRef] [PubMed]
36. Chivukula, V.K.; Levitt, M.R.; Clark, A.; Barbour, M.C.; Sansom, K.; Johnson, L.; Kelly, C.M.; Geindreau, C.; Rolland du Roscoat, S.; Kim, L.J.; et al. Reconstructing patient-specific cerebral aneurysm vasculature for in vitro investigations and treatment efficacy assessments. *J. Clin. Neurosci. Off. J. Neurosurg. Soc. Australas.* **2019**, *61*, 153–159. [CrossRef]
37. Paliwal, N.; Damiano, R.J.; Varble, N.A.; Tutino, V.M.; Dou, Z.; Siddiqui, A.H.; Meng, H. Methodology for Computational Fluid Dynamic Validation for Medical Use: Application to Intracranial Aneurysm. *J. Biomech. Eng.* **2017**, *139*, 1210041–12100410. [CrossRef]
38. Rossi de Aguiar, K.M.F.; Nascimento, M.V.; Faccioni, J.L.; Noeske, P.L.M.; Gätjen, L.; Rischka, K.; Rodrigues-Filho, U.P. Urethanes PDMS-based: Functional hybrid coatings for metallic dental implants. *Appl. Surf. Sci.* **2019**, *484*, 1128–1140. [CrossRef]
39. Tran, P.A.; Fox, K.; Tran, N. Novel hierarchical tantalum oxide-PDMS hybrid coating for medical implants: One pot synthesis, characterization and modulation of fibroblast proliferation. *J. Colloid Interface Sci.* **2017**, *485*, 106–115. [CrossRef] [PubMed]
40. Hijón, N.; Manzano, M.; Salinas, A.; Vallet-Regí, M. Bioactive CaO-SiO_2-PDMS coatings on Ti6Al4V substrates. *Chem. Mater.* **2005**, *17*, 1591–1596. [CrossRef]
41. Lee, D.S.; Kim, S.J.; Sohn, J.H.; Kim, I.G.; Kim, S.W.; Sohn, D.W.; Kim, J.H.; Choi, B. Biocompatibility of a pdms-coated micro-device: Bladder volume monitoring sensor. *Chin. J. Polym. Sci.* **2012**, *30*, 242–249. [CrossRef]
42. Tavakoli, S.; Nemati, S.; Kharaziha, M.; Akbari-Alavijeh, S. Embedding CuO nanoparticles in PDMS-SiO_2 coating to improve antibacterial characteristic and corrosion resistance. *Colloids Interface Sci. Commun.* **2019**, *28*, 20–28. [CrossRef]
43. Chen, S.; Jones, J.A.; Xu, Y.; Low, H.-Y.; Anderson, J.M.; Leong, K.W. Characterization of topographical effects on macrophage behavior in a foreign body response model. *Biomaterials* **2010**, *31*, 3479–3491. [CrossRef]
44. Guo, R.; Liu, J. Implantable liquid metal-based flexible neural microelectrode array and its application in recovering animal locomotion functions. *J. Micromech. Microeng.* **2017**, *27*, 104002. [CrossRef]
45. Hassler, C.; Boretius, T.; Stieglitz, T. Polymers for neural implants. *J. Polym. Sci. Part B Polym. Phys.* **2011**, *49*, 18–33. [CrossRef]
46. Wolf, M.P.; Salieb-Beugelaar, G.B.; Hunziker, P. PDMS with designer functionalities—Properties, modifications strategies, and applications. *Prog. Polym. Sci.* **2018**, *83*, 97–134. [CrossRef]
47. Zhao, J.; Sheadel, D.A.; Xue, W. Surface treatment of polymers for the fabrication of all-polymer MEMS devices. *Sens. Actuators A Phys.* **2012**, *187*, 43–49. [CrossRef]
48. Johnston, I.D.; McCluskey, D.K.; Tan, C.K.L.; Tracey, M.C. Mechanical characterization of bulk Sylgard 184 for microfluidics and microengineering. *J. Micromech. Microeng.* **2014**, *24*, 035017. [CrossRef]
49. Victor, A.; Ribeiro, J.; Araújo, F.F. Study of PDMS characterization and its applications in biomedicine: A review. *J. Mech. Eng. Biomech.* **2019**, *4*, 1–9. [CrossRef]
50. Cardoso, C.; Fernandes, C.S.; Lima, R.; Ribeiro, J. Biomechanical analysis of PDMS channels using different hyperelastic numerical constitutive models. *Mech. Res. Commun.* **2018**, *90*, 26–33. [CrossRef]
51. Pan, C.T.; Chen, Y.C.; Lin, P.-H.; Hsieh, C.C.; Hsu, F.T.; Lin, P.-H.; Chang, C.M.; Hsu, J.H.; Huang, J.C. Lens of controllable optical field with thin film metallic glasses for UV-LEDs. *Opt. Express* **2014**, *22*, 14411. [CrossRef]
52. Wang, B.; Liu, H.; Zhang, B.; Han, Y.; Shen, C.; Lin, Q.; Chen, H. Development of antibacterial and high light transmittance bulk materials: Incorporation and sustained release of hydrophobic or hydrophilic antibiotics. *Colloids Surf. B Biointerfaces* **2016**, *141*, 483–490. [CrossRef] [PubMed]
53. Pan, C.T.; Chen, Y.C.; Chen, Y.J.; Wang, W.C.; Yang, H.C.; Wu, H.C. Compound optical film using gray scale mask embedded with microvoids. *Adv. Condens. Matter Phys.* **2012**, *2012*, 942018. [CrossRef]
54. Riehle, N.; Thude, S.; Götz, T.; Kandelbauer, A.; Thanos, S.; Tovar, G.; Lorenz, G. Influence of PDMS molecular weight on transparency and mechanical properties of soft polysiloxane-urea-elastomers for intraocular lens application. *Eur. Polym. J.* **2018**, *101*, 190–201. [CrossRef]
55. Sales, F.; Souza, A.; Ariati, R.; Noronha, V.; Giovanetti, E.; Lima, R.; Ribeiro, J. Composite material of PDMS with interchangeable transmittance: Study of optical, mechanical properties and wettability. *J. Compos. Sci.* **2021**, *5*, 110. [CrossRef]
56. Mark, J.E. (Ed.) *Polymer Data Handbook*; Oxford University Press: Oxford, UK, 1999.
57. The Dow CompanyChemical. SYLGARD^{TM} 184 Silicone Elastomer Technical Datasheet. Silicone Elastomer Technical Data Sheet 2017. Available online: https://consumer.dow.com/en-us/document-viewer.html?ramdomVar=3835418757322904567&docPath=/documents/en-us/productdatasheet/11/11-31/11-3184-sylgard-184-elastomer.pdf (accessed on 20 August 2021).
58. Hong, J.; Lee, J.; Hong, C.; Shim, S. Effect of dispersion state of carbon nanotube on the thermal conductivity of poly(dimethyl siloxane) composites. *Curr. Appl. Phys.* **2010**, *10*, 359–363. [CrossRef]
59. Armani, D.; Liu, C.; Aluru, N. Re-Configurable Fluid Circuits by PDMS Elastomer Micromachining. In Proceedings of the Technical Digest IEEE International MEMS 99 Conference. Twelfth IEEE International Conference on Micro Electro Mechanical Systems (Cat. No.99CH36291), Orlando, FL, USA, 21 January 1999; pp. 222–227.
60. Müller, A.; Wapler, M.C.; Wallrabe, U. A quick and accurate method to determine the Poisson's ratio and the coefficient of thermal expansion of PDMS. *Soft Matter* **2019**, *15*, 779–784. [CrossRef] [PubMed]
61. Zhang, G.; Sun, Y.; Qian, B.; Gao, H.; Zuo, D. Experimental study on mechanical performance of polydimethylsiloxane (PDMS) at various temperatures. *Polym. Test.* **2020**, *90*, 106670. [CrossRef]

62. Gokaltun, A.; Yarmush, M.L.; Asatekin, A.; Usta, O.B. Recent advances in nonbiofouling PDMS surface modification strategies applicable to microfluidic technology. *Technology* **2017**, *5*, 1–12. [CrossRef] [PubMed]
63. GRIFFITHS, E. international critical tables of numerical data, physics, chemistry and technology. *Nature* **1927**, *119*, 735–738. [CrossRef]
64. Wu, M.H.; Urban, J.P.G.; Cui, Z.; Cui, Z.F. Development of PDMS microbioreactor with well-defined and homogenous culture environment for chondrocyte 3-D culture. *Biomed. Microdevices* **2006**, *8*, 331–340. [CrossRef] [PubMed]
65. Tan, S.H.; Nguyen, N.T.; Chua, Y.C.; Kang, T.G. Oxygen plasma treatment for reducing hydrophobicity of a sealed polydimethylsiloxane microchannel. *Biomicrofluidics* **2010**, *4*, 032204. [CrossRef]
66. Nakano, H.; Kakinoki, S.; Iwasaki, Y. Long-lasting hydrophilic surface generated on poly(dimethyl siloxane) with photoreactive zwitterionic polymers. *Colloids Surf. B Biointerfaces* **2021**, *205*, 111900. [CrossRef] [PubMed]
67. Lee, J.N.; Park, C.; Whitesides, G.M. Solvent compatibility of poly(dimethylsiloxane)-based microfluidic devices. *Anal. Chem.* **2003**, *75*, 6544–6554. [CrossRef] [PubMed]
68. Toepke, M.W.; Beebe, D.J. PDMS absorption of small molecules and consequences in microfluidic applications. *Lab Chip* **2006**, *6*, 1484–1486. [CrossRef] [PubMed]
69. Bodas, D.; Khan-Malek, C. Hydrophilization and hydrophobic recovery of PDMS by oxygen plasma and chemical treatment-An SEM investigation. *Sens. Actuators B Chem.* **2007**, *123*, 368–373. [CrossRef]
70. Makamba, H.; Kim, J.H.; Lim, K.; Park, N.; Hahn, J.H. Surface modification of poly(dimethylsiloxane) microchannels. *Electrophoresis* **2003**, *24*, 3607–3619. [CrossRef] [PubMed]
71. Zhou, J.; Khodakov, D.A.; Ellis, A.V.; Voelcker, N.H. Surface modification for PDMS-based microfluidic devices. *Electrophoresis* **2012**, *33*, 89–104. [CrossRef] [PubMed]
72. Hemmilä, S.; Cauich-Rodríguez, J.V.; Kreutzer, J.; Kallio, P. Rapid, simple, and cost-effective treatments to achieve long-term hydrophilic PDMS surfaces. *Appl. Surf. Sci.* **2012**, *258*, 9864–9875. [CrossRef]
73. Trantidou, T.; Elani, Y.; Parsons, E.; Ces, O. Hydrophilic surface modification of pdms for droplet microfluidics using a simple, quick, and robust method via PVA deposition. *Microsyst. Nanoeng.* **2017**, *3*, 16091. [CrossRef]
74. Yang, Y.; Kulangara, K.; Lam, R.T.S.; Dharmawan, R.; Leong, K.W. Effects of Topographical and mechanical property alterations induced by oxygen plasma modification on stem cell behavior. *ACS Nano* **2012**, *6*, 8591–8598. [CrossRef] [PubMed]
75. Berdichevsky, Y.; Khandurina, J.; Guttman, A.; Lo, Y.-H. UV/ozone modification of poly(dimethylsiloxane) microfluidic channels. *Sens. Actuators B Chem.* **2004**, *97*, 402–408. [CrossRef]
76. Hillborg, H.; Gedde, U.W. Hydrophobicity recovery of polydimethylsiloxane after exposure to corona discharges. *Polymer* **1998**, *39*, 1991–1998. [CrossRef]
77. Makamba, H.; Hsieh, Y.Y.; Sung, W.C.; Chen, S.H. Stable permanently hydrophilic protein-resistant thin-film coatings on poly(dimethylsiloxane) substrates by electrostatic self-assembly and chemical cross-linking. *Anal. Chem.* **2005**, *77*, 3971–3978. [CrossRef]
78. Boxshall, K.; Wu, M.-H.; Cui, Z.; Cui, Z.; Watts, J.F.; Baker, M.A. Simple surface treatments to modify protein adsorption and cell attachment properties within a poly(dimethylsiloxane) micro-bioreactor. *Surf. Interface Anal.* **2006**, *38*, 198–201. [CrossRef]
79. Blättler, T.M.; Pasche, S.; Textor, M.; Griesser, H.J. High salt stability and protein resistance of poly(L-lysine)-g-poly(ethylene glycol) copolymers covalently immobilized via aldehyde plasma polymer interlayers on inorganic and polymeric substrates. *Langmuir ACS J. Surf. Colloids* **2006**, *22*, 5760–5769. [CrossRef]
80. Xu, J.; Gleason, K.K. Conformal, amine-functionalized thin films by initiated chemical vapor deposition (iCVD) for hydrolytically stable microfluidic devices. *Chem. Mater.* **2010**, *22*, 1732–1738. [CrossRef]
81. Zhang, Z.; Feng, X.; Xu, F.; Liu, X.; Liu, B.F. "Click" chemistry-based surface modification of poly(dimethylsiloxane) for protein separation in a microfluidic chip. *Electrophoresis* **2010**, *31*, 3129–3136. [CrossRef] [PubMed]
82. Hu, S.; Ren, X.; Bachman, M.; Sims, C.E.; Li, G.P.; Allbritton, N.L. Surface-directed, graft polymerization within microfluidic channels. *Anal. Chem.* **2004**, *76*, 1865–1870. [CrossRef] [PubMed]
83. Ariati, R.; Sales, F.; Souza, A.; Lima, R.A.; Ribeiro, J. Polydimethylsiloxane composites characterization and its applications: A review. *Polymers* **2021**, *13*, 4258. [CrossRef]
84. Gökaltun, A.; Kang, Y.B.; Yarmush, M.L.; Usta, O.B.; Asatekin, A. Simple Surface Modification of Poly(dimethylsiloxane) via Surface Segregating Smart Polymers for Biomicrofluidics. *Sci. Rep.* **2019**, *9*, 97377. [CrossRef] [PubMed]
85. Souza, A.; Souza, M.S.; Pinho, D.; Agujetas, R.; Ferrera, C.; Lima, R.; Puga, H.; Ribeiro, J. 3D manufacturing of intracranial aneurysm biomodels for flow visualizations: Low cost fabrication processes. *Mech. Res. Commun.* **2020**, *107*, 103535. [CrossRef]
86. Santiago-Alvarado, A.; Cruz-Félix, A.S.; González-García, J.; Sánchez-López, O.; Mendoza-Jasso, A.J.; Hernández-Castillo, I. Polynomial fitting techniques applied to opto-mechanical properties of PDMS Sylgard 184 for given curing parameters. *Mater. Res. Express* **2020**, *7*, 45301. [CrossRef]
87. Khanafer, K.; Duprey, A.; Schlicht, M.; Berguer, R. Effects of strain rate, mixing ratio, and stress–strain definition on the mechanical behavior of the polydimethylsiloxane (PDMS) material as related to its biological applications. *Biomed. Microdevices* **2008**, *11*, 503. [CrossRef]
88. Prajzler, V.; Nekvindova, P.; Spirkova, J.; Novotny, M. The evaluation of the refractive indices of bulk and thick polydimethylsiloxane and polydimethyl-diphenylsiloxane elastomers by the prism coupling technique. *J. Mater. Sci. Mater. Electron.* **2017**, *28*, 7951–7961. [CrossRef]

89. Lamberti, A.; Marasso, S.L.; Cocuzza, M. PDMS membranes with tunable gas permeability for microfluidic applications. *RSC Adv.* **2014**, *4*, 61415–61419. [CrossRef]
90. Choi, J.S.; Piao, Y.; Seo, T.S. Fabrication of a circular PDMS microchannel for constructing a three-dimensional endothelial cell layer. *Bioprocess. Biosyst. Eng.* **2013**, *36*, 1871–1878. [CrossRef]
91. Yoo, B.Y.; Kim, B.H.; Lee, J.S.; Shin, B.H.; Kwon, H.; Koh, W.G.; Heo, C.Y. Dual surface modification of PDMS-based silicone implants to suppress capsular contracture. *Acta Biomater.* **2018**, *76*, 56–70. [CrossRef]
92. McMullan, D. Scanning electron microscopy 1928–1965. *Scanning* **1995**, *17*, 175–185. [CrossRef]
93. Wong, B.; Zhang, Z.; Handa, Y.P. High-precision gravimetric technique for determining the solubility and diffusivity of gases in polymers. *J. Polym. Sci. Part B Polym. Phys.* **1998**, *36*, 2025–2032. [CrossRef]
94. Goodwin, J.; Clark, C.; Deakes, J.; Burdon, D.; Lawrence, C. Clinical methods of goniometry: A comparative study. *Disabil. Rehabil.* **1992**, *14*, 10–15. [CrossRef] [PubMed]
95. Schuh, C.A. Nanoindentation studies of materials. *Mater. Today* **2006**, *9*, 32–40. [CrossRef]
96. Chong, H.; Lou, J.; Bogie, K.M.; Zorman, C.A.; Majerus, S.J.A. Vascular pressure-flow measurement using CB-PDMS flexible strain sensor. *IEEE Trans. Biomed. Circuits Syst.* **2019**, *13*, 1451–1461. [CrossRef] [PubMed]
97. Boyce, M.C.; Arruda, E.M. An experimental and anaiytical investigation of the large strain compressive and tensile response of glassy polymers. *Polym. Eng. Sci.* **1990**, *30*, 1288–1298. [CrossRef]
98. Fadley, C.S. X-ray photoelectron spectroscopy: Progress and perspectives. *J. Electron. Spectrosc. Relat. Phenom.* **2010**, *178–179*, 2–32. [CrossRef]
99. Koenig, J.L. *Fourier Transform Infrared Spectroscopy of Polymers BT—Spectroscopy: NMR, Fluorescence, FT-IR*; Springer: Berlin/Heidelberg, Germany, 1984; pp. 87–154.
100. Xia, Y.; Whitesides, G.M. Soft lithography. *Angew. Chem. Int. Ed.* **1998**, *37*, 550–575. [CrossRef]
101. Catarino, S.O.; Rodrigues, R.O.; Pinho, D.; Miranda, J.M.; Minas, G.; Lima, R. Blood cells separation and sorting techniques of passive microfluidic devices: From fabrication to applications. *Micromachines* **2019**, *10*, 593. [CrossRef]
102. Larson, R.G.; Rehg, T.J. Spin Coating BT. In *Liquid Film Coating: Scientific Principles and Their Technological Implications*; Kistler, S.F., Schweizer, P.M., Eds.; Springer: Dordrecht, The Netherlands, 1997; pp. 709–734.
103. Akther, F.; Yakob, S.B.; Nguyen, N.-T.; Ta, H.T. Surface modification techniques for endothelial cell seeding in PDMS microfluidic devices. *Biosensors* **2020**, *10*, 182. [CrossRef]
104. Morra, M.; Occhiello, E.; Marola, R.; Garbassi, F.; Humphrey, P.; Johnson, D. On the aging of oxygen plasma-treated polydimethylsiloxane surfaces. *J. Colloid Interface Sci.* **1990**, *137*, 11–24. [CrossRef]
105. Xiong, L.; Chen, P.; Zhou, Q. Adhesion promotion between PDMS and glass by oxygen plasma pre-treatment. *J. Adhes. Sci. Technol.* **2014**, *28*, 1046–1054. [CrossRef]
106. Liu, J.; Yao, Y.; Li, X.; Zhang, Z. Fabrication of advanced polydimethylsiloxane-based functional materials: Bulk modifications and surface functionalizations. *Chem. Eng. J.* **2021**, *408*, 127262. [CrossRef]
107. Shin, S.; Kim, N.; Hong, J.W. Comparison of surface modification techniques on polydimethylsiloxane to prevent protein adsorption. *BioChip J.* **2018**, *12*, 123–127. [CrossRef]
108. Zhao, Y.; Wen, J.; Ge, Y.; Zhang, X.; Shi, H.; Yang, K.; Gao, X.; Shi, S.; Gong, Y. Fabrication of stable biomimetic coating on PDMS surface: Cooperativity of multivalent interactions. *Appl. Surf. Sci.* **2019**, *469*, 720–730. [CrossRef]
109. Jinia, A.J.; Sunbul, N.B.; Meert, C.A.; Miller, C.A.; Clarke, S.D.; Kearfott, K.J.; Matuszak, M.M.; Pozzi, S.A. Review of sterilization techniques for medical and personal protective equipment contaminated with SARS-CoV-2. *IEEE Access* **2020**, *8*, 111347–111354. [CrossRef] [PubMed]
110. Linke, B. Sterilization Methods and Impact on Electronics in Medical Devices. 2011. Available online: https://www.eetimes.com/document.asp?doc_id=1278906 (accessed on 20 September 2021).
111. Harrington, R.E.; Guda, T.; Lambert, B.; Martin, J. *Sterilization and Disinfection of Biomaterials for Medical Devices*, 4th ed.; Elsevier: Amsterdam, The Netherlands, 2020.
112. Tipnis, N.P.; Burgess, D.J. Sterilization of implantable polymer-based medical devices: A review. *Int. J. Pharm.* **2018**, *544*, 455–460. [CrossRef]
113. Rogers, W.J. *Sterilisation Techniques for Polymers*; Elsevier Masson SAS: Issy-les-Moulineaux, France, 2012.
114. Dai, Z.; Ronholm, J.; Tian, Y.; Sethi, B.; Cao, X. Sterilization techniques for biodegradable scaffolds in tissue engineering applications. *J. Tissue Eng.* **2016**, *7*, 2041731416648810. [CrossRef]
115. Zhai, H.; Li, J.; Chen, Z.; Su, Z.; Liu, Z.; Yu, X. A glass/PDMS electrophoresis microchip embedded with molecular imprinting SPE monolith for contactless conductivity detection. *Microchem. J.* **2014**, *114*, 223–228. [CrossRef]
116. Schöning, M.J.; Jacobs, M.; Muck, A.; Knobbe, D.T.; Wang, J.; Chatrathi, M.; Spillmann, S. Amperometric PDMS/glass capillary electrophoresis-based biosensor microchip for catechol and dopamine detection. *Sens. Actuators B Chem.* **2005**, *108*, 688–694. [CrossRef]
117. Xu, K.; Clark, C.P.; Poe, B.L.; Lounsbury, J.A.; Nilsson, J.; Laurell, T.; Landers, J.P. Isolation of a low number of sperm cells from female DNA in a glass-PDMS-Glass microchip via bead-assisted acoustic differential extraction. *Anal. Chem.* **2019**, *91*, 2186–2191. [CrossRef]
118. Hong, J.W.; Fujii, T.; Seki, M.; Yamamoto, T.; Endo, I. Integration of gene amplification and capillary gel electrophoresis on a polydimethylsiloxane-glass hybrid microchip. *Electrophoresis* **2001**, *22*, 328–333. [CrossRef]

119. Xia, Y.-M.; Hua, Z.-S.; Srivannavit, O.; Ozel, A.; Gulari, E. Minimizing the surface effect of PDMS-glass microchip on polymerase chain reaction by dynamic polymer Passivation. *J. Chem. Technol. Biotechnol.* **2007**, *82*, 33–38. [CrossRef]
120. Niu, Z.Q.; Chen, W.Y.; Shao, S.Y.; Jia, X.Y.; Zhang, W.P. DNA amplification on a PDMS-glass hybrid microchip. *J. Micromech. Microeng.* **2006**, *16*, 425–433. [CrossRef]
121. Nandi, P.; Desai, D.P.; Lunte, S.M. Development of a PDMS-based microchip electrophoresis device for continuous online in vivo monitoring of microdialysis samples. *Electrophoresis* **2010**, *31*, 1414–1422. [CrossRef]
122. Ha, B.H.; Destgeer, G.; Park, J.; Jung, J.H.; Sung, H.J. Generation of complex, dynamic temperature gradients in a disposable microchip. *Phys. Procedia* **2015**, *70*, 38–41. [CrossRef]
123. Rodrigues, R.O.; Pinho, D.; Bento, D.; Lima, R.; Ribeiro, J. Wall expansion assessment of an intracranial aneurysm model by a 3D Digital Image Correlation System. *Measurement* **2016**, *88*, 262–270. [CrossRef]
124. Fiddes, L.K.; Raz, N.; Srigunapalan, S.; Tumarkan, E.; Simmons, C.A.; Wheeler, A.R.; Kumacheva, E. A circular cross-section PDMS microfluidics system for replication of cardiovascular flow conditions. *Biomaterials* **2010**, *31*, 3459–3464. [CrossRef]
125. Siddique, A.; Pause, I.; Narayan, S.; Kruse, L.; Stark, R.W. Endothelialization of PDMS-based microfluidic devices under high shear stress conditions. *Colloids Surf. B Biointerfaces* **2021**, *197*, 111394. [CrossRef]
126. Lima, R.; Wada, S.; Tanaka, S.; Takeda, M.; Ishikawa, T.; Tsubota, K.I.; Imai, Y.; Yamaguchi, T. In vitro blood flow in a rectangular PDMS microchannel: Experimental observations using a confocal micro-PIV system. *Biomed. Microdevices* **2008**, *10*, 153–167. [CrossRef]
127. Lima, R.; Oliveira, M.S.N.; Ishikawa, T.; Kaji, H.; Tanaka, S.; Nishizawa, M.; Yamaguchi, T. Axisymmetric polydimethysiloxane microchannels for in vitro hemodynamic studies. *Biofabrication* **2009**, *1*, 35005. [CrossRef] [PubMed]
128. Nguyen, T.Q.; Park, W.-T. Fabrication method of multi-depth circular microchannels for investigating arterial thrombosis-on-a-chip. *Sens. Actuators B Chem.* **2020**, *321*, 128590. [CrossRef]
129. Agrawal, S.; Paknikar, K.; Bodas, D. Development of immunosensor using magnetic nanoparticles and circular microchannels in PDMS. *Microelectron. Eng.* **2014**, *115*, 66–69. [CrossRef]
130. Morarka, A.; Agrawal, S.; Kale, S.; Kale, A.; Ogale, S.; Paknikar, K.; Bodas, D. Quantum dot based immunosensor using 3D circular microchannels fabricated in PDMS. *Biosens. Bioelectron.* **2011**, *26*, 3050–3053. [CrossRef] [PubMed]
131. Fujiwara, H.; Ishikawa, T.; Lima, R.; Matsuki, N.; Imai, Y.; Kaji, H.; Nishizawa, M.; Yamaguchi, T. Red blood cell motions in high-hematocrit blood flowing through a stenosed microchannel. *J. Biomech.* **2009**, *42*, 838–843. [CrossRef] [PubMed]
132. Zeng, N.F.; Ristenpart, W.D. Mechanical response of red blood cells entering a constriction. *Biomicrofluidics* **2014**, *8*, 64123. [CrossRef] [PubMed]
133. Faustino, V.; Catarino, S.; Pinho, D.; Lima, R.; Minas, G. A Passive microfluidic device based on crossflow filtration for cell separation measurements: A spectrophotometric characterization. *Biosensors* **2018**, *8*, 125. [CrossRef]
134. Zhao, R.; Antaki, J.F.; Naik, T.; Bachman, T.N.; Kameneva, M.V.; Wu, Z.J. Microscopic investigation of erythrocyte deformation dynamics. *Biorheology* **2006**, *43*, 747–765.
135. Lee, S.S.; Yim, Y.; Ahn, K.H.; Lee, S.J. Extensional flow-based assessment of red blood cell deformability using hyperbolic converging microchannel. *Biomed. Microdevices* **2009**, *11*, 1021–1027. [CrossRef]
136. Pinho, D.; Yaginuma, T.; Lima, R. A microfluidic device for partial cell separation and deformability assessment. *BioChip J.* **2013**, *7*, 367–374. [CrossRef]
137. Bento, D.; Sousa, L.; Yaginuma, T.; Garcia, V.; Lima, R.; Miranda, J.M. Microbubble moving in blood flow in microchannels: Effect on the cell-free layer and cell local concentration. *Biomed. Microdevices* **2017**, *19*, 6. [CrossRef]
138. Bento, D.; Rodrigues, R.; Faustino, V.; Pinho, D.; Fernandes, C.; Pereira, A.; Garcia, V.; Miranda, J.; Lima, R. Deformation of red blood cells, air bubbles, and droplets in microfluidic devices: Flow visualizations and measurements. *Micromachines* **2018**, *9*, 151. [CrossRef]
139. Shelby, J.P.; White, J.; Ganesan, K.; Rathod, P.K.; Chiu, D.T. A microfluidic model for single-cell capillary obstruction by Plasmodium falciparum-infected erythrocytes. *Proc. Natl. Acad. Sci. USA* **2003**, *100*, 14618–14622. [CrossRef]
140. Boas, L.V.; Faustino, V.; Lima, R.; Miranda, J.M.; Minas, G.; Fernandes, C.S.V.; Catarino, S.O. Assessment of the deformability and velocity of healthy and artificially impaired red blood cells in narrow polydimethylsiloxane (PDMS) microchannels. *Micromachines* **2018**, *9*, 384. [CrossRef]
141. Hou, H.W.; Li, Q.S.; Lee, G.Y.H.; Kumar, A.P.; Ong, C.N.; Lim, C.T. Deformability study of breast cancer cells using microfluidics. *Biomed. Microdevices* **2009**, *11*, 557–564. [CrossRef] [PubMed]
142. Faustino, V.; Pinho, D.; Yaginuma, T.; Calhelha, R.C.; Ferreira, I.C.F.R.; Lima, R. Extensional flow-based microfluidic device: Deformability assessment of red blood cells in contact with tumor cells. *BioChip J.* **2014**, *8*, 42–47. [CrossRef]
143. Faustino, V.; Rodrigues, R.O.; Pinho, D.; Costa, E.; Santos-Silva, A.; Miranda, V.; Amaral, J.S.; Lima, R. A microfluidic deformability assessment of pathological red blood cells flowing in a hyperbolic converging microchannel. *Micromachines* **2019**, *10*, 645. [CrossRef] [PubMed]
144. Rodrigues, R.O.; Pinho, D.; Faustino, V.; Lima, R. A simple microfluidic device for the deformability assessment of blood cells in a continuous flow. *Biomed. Microdevices* **2015**, *17*, 108. [CrossRef]
145. Rodrigues, R.O.; Bañobre-López, M.; Gallo, J.; Tavares, P.B.; Silva, A.M.T.; Lima, R.; Gomes, H.T. Haemocompatibility of iron oxide nanoparticles synthesized for theranostic applications: A high-sensitivity microfluidic tool. *J. Nanopart. Res.* **2016**, *18*, 194. [CrossRef]

146. Bento, D.; Fernandes, C.S.; Miranda, J.M.; Lima, R. In vitro blood flow visualizations and cell-free layer (CFL) measurements in a microchannel network. *Exp. Therm. Fluid Sci.* **2019**, *109*, 109847. [CrossRef]
147. Pinto, E.; Faustino, V.; Rodrigues, R.O.; Pinho, D.; Garcia, V.; Miranda, J.M.; Lima, R. A rapid and low-cost nonlithographic method to fabricate biomedical microdevices for blood flow analysis. *Micromachines* **2015**, *6*, 121–135. [CrossRef]
148. Bento, D.; Lopes, S.; Maia, I.; Lima, R.; Miranda, J.M. Bubbles moving in blood flow in a microchannel network: The effect on the local hematocrit. *Micromachines* **2020**, *11*, 344. [CrossRef]
149. Bento, D.; Pereira, A.I.; Lima, J.; Miranda, J.M.; Lima, R. Cell-free layer measurements of in vitro blood flow in a microfluidic network: An automatic and manual approach. *Comput. Methods Biomech. Biomed. Eng. Imaging Vis.* **2018**, *6*, 629–637. [CrossRef]
150. Leble, V.; Lima, R.; Dias, R.; Fernandes, C.; Ishikawa, T.; Imai, Y.; Yamaguchi, T. Asymmetry of red blood cell motions in a microchannel with a diverging and converging bifurcation. *Biomicrofluidics* **2011**, *5*, 044120. [CrossRef]
151. Ishikawa, T.; Fujiwara, H.; Matsuki, N.; Yoshimoto, T.; Imai, Y.; Ueno, H.; Yamaguchi, T. Asymmetry of blood flow and cancer cell adhesion in a microchannel with symmetric bifurcation and confluence. *Biomed. Microdevices* **2011**, *13*, 159–167. [CrossRef]
152. Faustino, V.; Catarino, S.O.; Lima, R.; Minas, G. Biomedical microfluidic devices by using low-cost fabrication techniques: A review. *J. Biomech.* **2016**, *49*, 2280–2292. [CrossRef] [PubMed]
153. Sadek, S.H.; Rubio, M.; Lima, R.; Vega, E.J. Blood Particulate Analogue Fluids: A Review. *Materials* **2021**, *14*, 2451. [CrossRef] [PubMed]
154. Carvalho, V.; Maia, I.; Souza, A.; Ribeiro, J.; Costa, P.; Puga, H.; Teixeira, S.; Lima, R.A. In vitro biomodels in stenotic arteries to perform blood analogues flow visualizations and measurements: A review. *Open Biomed. Eng. J.* **2021**, *14*, 87–102. [CrossRef]
155. Carvalho, V.; Gonçalves, I.M.; Souza, A.; Souza, M.S.; Bento, D.; Ribeiro, J.E.; Lima, R.; Pinho, D. Manual and automatic image analysis segmentation methods for blood flow studies in microchannels. *Micromachines* **2021**, *12*, 317. [CrossRef]
156. Pinho, D.; Carvalho, V.; Gonçalves, I.M.; Teixeira, S.; Lima, R. Visualization and measurements of blood cells flowing in microfluidic systems and blood rheology: A personalized medicine perspective. *J. Pers. Med.* **2020**, *10*, 249. [CrossRef]
157. Calejo, J.; Pinho, D.; Galindo-Rosales, F.J.; Lima, R.; Campo-Deaño, L. Particulate blood analogues reproducing the erythrocytes cell-free layer in a microfluidic device containing a hyperbolic contraction. *Micromachines* **2016**, *7*, 4. [CrossRef] [PubMed]
158. Sun, H.; Björnmalm, M.; Cui, J.; Wong, E.H.H.; Dai, Y.; Dai, Q.; Qiao, G.G.; Caruso, F. Structure governs the deformability of polymer particles in a microfluidic blood capillary model. *ACS Macro Lett.* **2015**, *4*, 1205–1209. [CrossRef]
159. Maruyama, O.; Yamane, T.; Nishida, M.; Aouidef, A.; Tsutsui, T.; Jikuya, T.; Masuzawa, T. Fractural characteristic evaluation of a microcapsule suspension using a rotational shear stressor. *ASAIO J.* **2002**, *48*, 365–373. [CrossRef] [PubMed]
160. Carvalho, D.A.M.; Rodrigues, A.R.O.; Faustino, V.; Pinho, D.; Castanheira, E.M.S.; Lima, R. Microfluidic deformability study of an innovative blood analogue fluid based on giant unilamellar vesicles. *J. Funct. Biomater.* **2018**, *9*, 70. [CrossRef] [PubMed]
161. Lima, R.; Vega, E.J.; Moita, A.S.; Miranda, J.M.; Pinho, D.; Moreira, A.L.N. Fast, flexible and low-cost multiphase blood analogue for biomedical and energy applications. *Exp. Fluids* **2020**, *61*, 231. [CrossRef]
162. Merkel, T.J.; Jones, S.W.; Herlihy, K.P.; Kersey, F.R.; Shields, A.R.; Napier, M.; Luft, J.C.; Wu, H.; Zamboni, W.C.; Wang, A.Z.; et al. Using mechanobiological mimicry of red blood cells to extend circulation times of hydrogel microparticles. *Proc. Natl. Acad. Sci. USA* **2011**, *108*, 586–591. [CrossRef]
163. Vilanova, N.; Rodríguez-Abreu, C.; Fernández-Nieves, A.; Solans, C. Fabrication of novel silicone capsules with tunable mechanical properties by microfluidic techniques. *ACS Appl. Mater. Interfaces* **2013**, *5*, 5247–5252. [CrossRef]
164. Cui, J.; Björnmalm, M.; Liang, K.; Xu, C.; Best, J.P.; Zhang, X.; Caruso, F. Super-soft hydrogel particles with tunable elasticity in a microfluidic blood capillary model. *Adv. Mater.* **2014**, *26*, 7295–7299. [CrossRef]
165. She, S.; Li, Q.; Shan, B.; Tong, W.; Gao, C. Fabrication of red-blood-cell-like polyelectrolyte microcapsules and their deformation and recovery behavior through a microcapillary. *Adv. Mater.* **2013**, *25*, 5814–5818. [CrossRef] [PubMed]
166. Pinho, D.; Campo-Deaño, L.; Lima, R.; Pinho, F.T. In vitro particulate analogue fluids for experimental studies of rheological and hemorheological behavior of glucose-rich RBC suspensions. *Biomicrofluidics* **2017**, *11*, 54105. [CrossRef]
167. Muñoz-Sánchez, B.N.; Silva, S.F.; Pinho, D.; Vega, E.J.; Lima, R. Generation of micro-sized PDMS particles by a flow focusing technique for biomicrofluidics applications. *Biomicrofluidics* **2016**, *10*, 014122. [CrossRef]
168. Anes, C.F.; Pinho, D.; Muñoz-Sánchez, B.N.; Vega, E.J.; Lima, R. Shrinkage and colour in the production of micro-sized PDMS particles for microfluidic applications. *J. Micromech. Microeng.* **2018**, *28*, 75002. [CrossRef]
169. Pinho, D.; Muñoz-Sánchez, B.N.; Anes, C.F.; Vega, E.J.; Lima, R. Flexible PDMS microparticles to mimic RBCs in blood particulate analogue fluids. *Mech. Res. Commun.* **2019**, *100*, 18–20. [CrossRef]
170. Choi, Y.H.; Chung, K.H.; Hong, H.B.; Lee, W.S. Production of PDMS microparticles by emulsification of two phases and their potential biological application. *Int. J. Polym. Mater. Polym. Biomater.* **2018**, *67*, 686–692. [CrossRef]
171. López, M.; Rubio, M.; Sadek, S.H.; Vega, E.J. A simple emulsification technique for the production of micro-sized flexible powder of polydimethylsiloxane (PDMS). *Powder Technol.* **2020**, *366*, 610–616. [CrossRef]
172. Carneiro, J.; Lima, R.; Campos, J.B.L.M.; Miranda, J.M. A microparticle blood analogue suspension matching blood rheology. *Soft Matter* **2021**, *17*, 3963–3974. [CrossRef] [PubMed]

173. Lima, R.; Vega, E.J.; Cardoso, V.F.; Minas, G.; Montanero, J.M. *Magnetic PDMS Microparticles for Biomedical and Energy Applications BT—VipIMAGE*; Tavares, J.M.R.S., Natal Jorge, R.M., Eds.; Springer International Publishing: Cham, Switzerland, 2019; pp. 578–584.
174. Barthes, J.; Lagarrigue, P.; Riabov, V.; Lutzweiller, G.; Kirsch, J.; Muller, C.; Courtial, E.-J.; Marquette, C.; Projetti, F.; Kzhyskowska, J.; et al. Biofunctionalization of 3D-printed silicone implants with immunomodulatory hydrogels for controlling the innate immune response: An in vivo model of tracheal defect repair. *Biomaterials* **2021**, *268*, 120549. [CrossRef]

Journal of
Functional Biomaterials

MDPI

Article

Physical Characterization and In Vitro Toxicity Test of PDMS Synthesized from Low-Grade D4 Monomer as a Vitreous Substitute in the Human Eyes

Diba Grace Auliya [1,*], Soni Setiadji [2,3], Fitrilawati Fitrilawati [1] and Risdiana Risdiana [1,*]

1 Department of Physics, Faculty of Mathematics and Natural Sciences, Universitas Padjadjaran, Jl. Raya Bandung-Sumedang km 21 Jatinangor, Sumedang 45363, Indonesia; fitrilawati@phys.unpad.ac.id
2 Department of Chemistry, Faculty of Mathematics and Natural Sciences, Universitas Padjadjaran, Jl. Raya Bandung-Sumedang km 21 Jatinangor, Sumedang 45363, Indonesia; s.setiadji@uinsgd.ac.id
3 Department of Chemistry, Faculty of Sciences and Technology, UIN Sunan Gunung Djati Bandung, Jl. A.H. Nasution No. 105, Cibiru, Bandung 40614, Indonesia
* Correspondence: diba15001@mail.unpad.ac.id (D.G.A.); risdiana@phys.unpad.ac.id (R.R.)

Abstract: Polydimethylsiloxane (PDMS) is one of the most superior materials and has been used as a substitute for vitreous humor in the human eye. In previous research, we have succeeded in producing PDMS with low and medium viscosity using octamethylcyclotetrasiloxane (D4) monomer with a low grade of 96%. Both have good physical properties and are comparable to commercial product PDMS and PDMS synthesized using D4 monomer with a high grade of 98%. An improvement of the synthesis process is needed to ensure that PDMS synthesized from a low-grade D4 monomer under specific synthesis conditions can repeatedly produce high-quality PDMS. Apart from good physical properties, the PDMS as a substitute for vitreous humor must also be safe and not cause other disturbances to the eyes. Here, we reported the process of synthesizing and characterizing the physical properties of low- and medium-viscosity PDMS using a low-grade D4 monomer. We also reported for the first time the in vitro toxicity test using the Hen's Egg Test Chorioallantoic Membrane (HET-CAM) test method. We have succeeded in obtaining PDMS with viscosities of 1.15 Pa.s, 1.17 Pa.s, and 1.81 Pa.s. All samples have good physical properties such as refractive index, surface tension, and functional groups that are similar to commercial PDMS. The HET-CAM test results showed that all samples did not show signs of irritation indicating that samples were non-toxic. From the results of this study, it can be concluded that PDMS synthesized from a low-grade D4 monomer under specific synthesis conditions by the ROP method is very safe and has the potential to be developed as a substitute for vitreous humor in human eyes.

Keywords: HET-CAM; high surface tension; low-grade D4; polydimethylsiloxane; toxicity

1. Introduction

Polydimethylsiloxane (PDMS) is one of the most superior materials and is commonly used to replace the vitreous humor damaged in the human eye through vitreoretinal surgery. Low-viscosity PDMS are preferred in vitreoretinal surgery because they are easier and faster to inject and expel than high-viscosity PDMS. However, PDMS with low viscosity also has a weakness in the form of a higher possibility of emulsification than PDMS with high viscosity [1]. Impurities are one of the main reasons for emulsification [2,3]. These impurities are thought to cause ocular toxicity by spreading to the surrounding eye tissue. Medium-viscosity PDMS is present as a new type of PDMS that offers advantages where the material is easier to inject than PDMS with high viscosity and has a lower emulsification tendency than other types of PDMS [4].

PDMS is synthesized from the octamethylcyclotetrasiloxane (D4) monomer. According to the European Chemical Agency (ECHA), D4 monomer is a toxic substance [5]. Several studies reported that D4 monomer has been shown to cause severe inflammation in rabbit and human eyes by penetrating ocular tissue [5–7]. D4 monomer caused acute ocular toxicity in the form of severe corneal edema and opacification [6]. Therefore, products produced from D4 monomer may contain hazardous and toxic substances. PDMS as a vitreous substitute will be used in human eyes and it must be non-toxic. For this reason, it is necessary to test the toxicity of PDMS to ensure the safety of using PDMS as a substitute for vitreous.

Various tests to replace rabbits in detecting potential chemical irritants have been developed [8]. One of them is the chicken egg chorioallantoic membrane (CAM) developed to detect a chemical test called the Hen's Egg Test Chorioallantoic Membrane (HET-CAM) assay [9]. The existence of the HET-CAM test can be an initial test for eye irritation tests so that the use of animals in toxicological tests can be reduced [8]. CAM is a complete set of arteries, capillaries, and veins. CAM is technically easy to learn. Changes in vascular injury in the form of bleeding, lysis, and coagulation respond to certain irritant chemicals [10]. Previous studies have reported a good correlation between in vitro tests with HET-CAM and in vivo tests with the Draize method. The HET-CAM test has advantages over other tests due to its speed, simplicity, convenience, and relatively low cost [11].

Previous studies have reported that the synthesis of PDMS using low-grade D4 monomer has been successfully carried out by the ring-opening polymerization (ROP) method and produced low-viscosity PDMS with properties similar to commercial products and PDMS synthesized using high-grade D4 monomer [12]. Medium viscosity PDMS has also been successfully synthesized using the same monomer by optimizing the synthesis parameters [13]. However, it is necessary to improve the synthesis process by repeating the synthesis to ensure that the PDMS synthesized from a low-grade D4 monomer under these synthesis conditions has high-quality PDMS. In addition, further information, especially the level of toxicity of PDMS synthesized from low-grade D4 monomer, is also unknown. The test is essential considering that PDMS will be used in the human body to guarantee its safety. Our study focuses on how to produce high-quality PDMS from low-grade D4 monomer in the terms of its viscosity, refractive index, surface tension, material content, and safety suitable for use as a vitreous substitute. The synthesis process is an important part of achieving this goal. Here, we reported the synthesis and its physical characterization of low and medium-viscosity PDMS using a low-grade D4 monomer. We also reported for the first time the in vitro toxicity test of these PDMS samples using the HET-CAM test method to obtain information safety and toxicity level of all PDMS samples.

2. Materials and Methods

The low viscosity and medium viscosity of PDMS were synthesized by the ROP method. ROP is a chain-growth polymerization with the end of the polymer chain acts as a reactive center. The mechanism of ROP in the synthesis of PDMS is based on the cleavage of Si-O in the monomer used. D4 monomer with a high grade of 98% was also used as a comparison for low viscosity PDMS using D4 monomer with a low grade of 96%. The ROP mechanism requires an initiator assisted by providing heat treatment.

The ROP process for the synthesis of PDMS consists of initiation, propagation, and termination. The use of an initiator in the form of potassium hydroxide (KOH) which is a strong base will result in an anionic ROP mechanism. A number of KOH with a certain concentration will initiate and form an anion, which represents the active center in the propagation reaction. At the initiation process, the OH ion from KOH donates a pair of electrons to one of the silicon atoms (Si) of the D4 siloxane and binds. As the result, the electron pairs that form siloxane bonds in the monomer ring break from the cyclic chain into a linear chain. The oxygen that gets a pair of electrons with a negative charge will bind to the molecule in the second cyclic monomer, and so on. Termination of the chain occurs due to the use of the end-capping agents, which in this study used disiloxane in the form of hexamethyldisiloxane (MM).

The low viscosity of PDMS using a low grade of D4 monomer coded as Sample A, while low viscosity of PDMS using a high grade of D4 monomer coded as Sample B. Medium viscosity of PDMS coded as Sample C. The synthesis begins with setting the synthesis temperature, then mixing 7.8 mL D4 and 3 mL MM (for volume ratio of D4:MM = 26:10) or 8.9 mL D4 and 1.9 mL MM (for volume ratio of D4:MM = 46:10). After that, a KOH solution with a certain concentration is added. The mixture of these materials is stirred for the specified time to form a gel. Synthesis conditions of all samples are listed in Table 1.

Table 1. Synthesis parameters of the samples.

Condition	A	B	C
Purity of D4 (%)	96	98	96
Ratio of D4:MM	26:10	26:10	46:10
Synthesis Temperature (°C)	190	150	190
KOH Concentration (M)	2	0.6	2
Time of Polymerization (minute)	50	16	50

The purification process was carried out by diluting the sample with chloroform. After that, milli-Q water was added. Then the samples are stored until the mixture separates into a liquid and gel phase. Both of them are separated. The pH of the liquid solution was checked to produce a neutral pH value. This purification process was repeated three times. Furthermore, stirring by heating is carried out to remove the chloroform.

PDMS samples were characterized to measure viscosity, refractive index, surface tension, and detect a functional group of the sample. The HET-CAM toxicity test was carried out using seven-day-old fertile white leghorn eggs weighing between 50–60 g. Eggs were tested for test materials (samples A, B, and, C), positive control (1% sodium dodecyl sulfate (SDS)), and negative control (0.9% NaCl). Each substance was tested on three eggs. After that, the eggs will be incubated at a temperature of 38.3 ± 0.2 °C for 10 days. The eggs that have been selected would be disinfected. The shells were removed to expose the membrane of the egg. The membranes were tested against the test material, positive control, and negative control. Observations were made for 300 s by recording the appearance time of each observed endpoint at time intervals of 0 s, 10 s, 30 s, 60 s, 180 s, and 300 s. The evaluation was carried out based on the percentage of the occurrence of the endpoint in the form of hemorrhage, lysis, and coagulation and given a value or scoring as shown in Table 2. The score value is adjusted to the standard reported by the Interagency Coordinating Committee on the Validation of Alternative Methods (ICCVAM) [14].

Table 2. Scoring scheme to irritant tests with the Hen's Egg Test Chorioallantoic Membrane (HET-CAM) method.

Area of the Endpoint (%)	Score
0–10	0
10–30	1
30–60	2
60–100	3

3. Results

3.1. PDMS Properties

Table 3 shows the characteristics of viscosity (η), yield, refractive index (n), additional diopters, and surface tension (γ) of all samples. The viscosity value of sample A has a slight difference from sample B. However, both of them are still in the range of low viscosity type. Sample A and sample B are categorized as low viscosity. Meanwhile, sample C has a medium viscosity. All samples have a transparent appearance, as shown in Figure 1.

Table 3. Polydimethylsiloxane (PDMS) characteristics of η, yield, n, additional diopters, and γ.

Sample	η (Pa.s)	Yield (%)	n	Additional Diopters	γ (mN/m)
A	1.15	67.37	1.4040	3.410	21
B	1.17	54.59	1.3993	3.179	19
C	1.81	71.31	1.4048	3.449	21.5

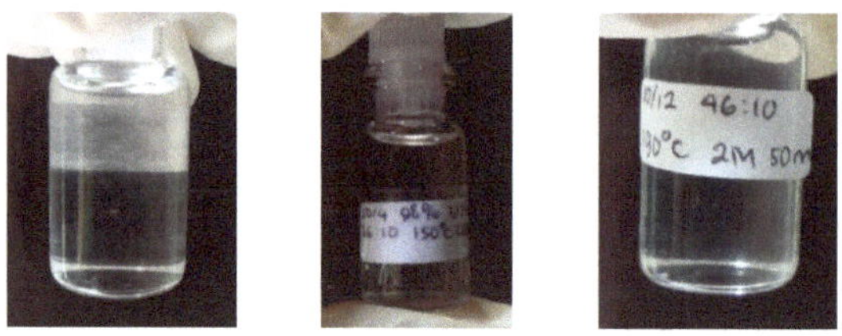

(a) (b) (c)

Figure 1. PDMS of (**a**) sample A; (**b**) sample B; and (**c**) sample C.

3.2. IR Spectra of PDMS

Functional groups of all samples have been identified and listed in Table 4. Compared with commercial PDMS, all of the samples have slight differences in wavenumber. Nevertheless, the transmittance peaks of all samples show the same spectra and indicate that all of them have the same functional group as the commercial. The infrared (IR) spectra of the samples and commercial are shown in Figure 2.

Table 4. Functional group of all samples and commercial PDMS.

No	Functional Group	Wavenumber (cm^{-1})			
		Commercial [13]	A	B	C
1	Si-O-Si	500, 703	523, 700	551, 695	519, 701
2	Si-C stretching and CH_3 rocking	792, 871	803, 864	807, 863	799, 864
3	Si-O-Si stretching	1112, 1023	1023 1099	1023, 1075	1019, 1093
4	CH_3 symmetric deformation of Si-CH_3	1263	1261	1261	1260
5	CH_3 asymmetric deformation of Si-CH_3	1412	1413	1412	1412
6	OH	1600, 3643–3828	1600, 3595–3823	1600, 3643–3828	1602, 3593–3828
7	Si-C	1945, 2052	1945, 2052	1945, 2052	1945, 2052
8	CH	2500, 2663	2500, 2663	2500, 2663	2500, 2663
9	CH stretching of CH_3	2906, 2972	2906, 2963	2906, 2964	2906, 2964

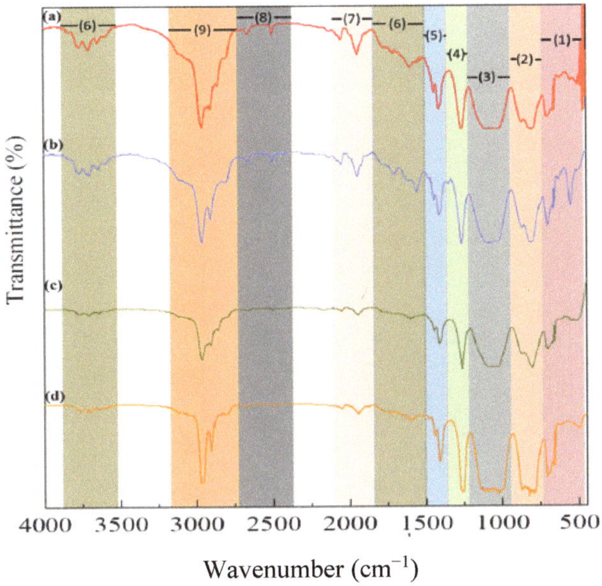

Figure 2. IR spectra of (**a**) sample A; (**b**) sample B; (**c**) sample C; and (**d**) commercial.

3.3. In Vitro Toxicity Test

The result scores of all samples and reference substances are shown in Figure 3. A significant change happens in the positive control group from 10 s until 300 s. The samples, comparison substances, and negative control did not show a significant change. More complete scoring results for hemorrhage, lysis and coagulation can be seen in Supplementary Materials of Figures S1–S3. Micro images of a blood vessel in the HET-CAM test are shown in Figure 4. Blood vessel damage was seen significantly in the positive control group, while the samples, reference substances, and negative control, did not damage the blood vessel. Detail figures of blood vessel micro images can be seen in Supplementary Materials of Figures S4–S8.

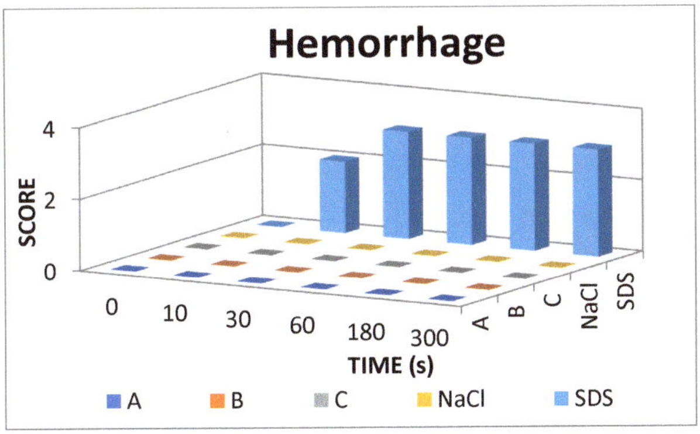

(**a**)

Figure 3. *Cont.*

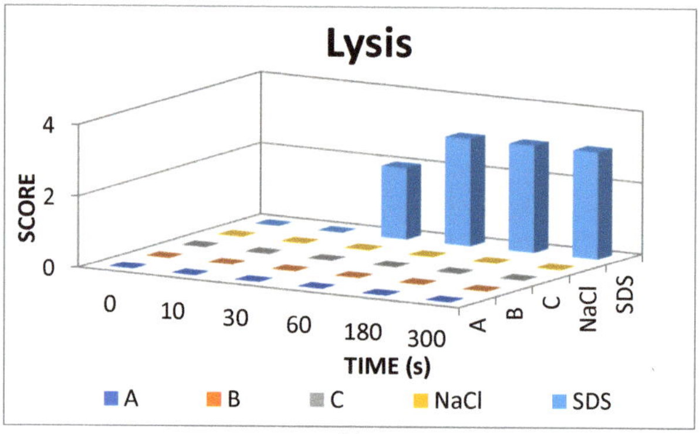

(b)

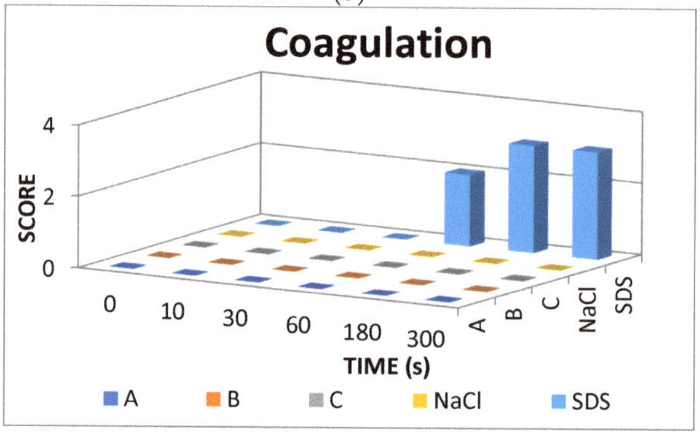

(c)

Figure 3. Endpoint score of (**a**) hemorrhage, (**b**) lysis, and (**c**) coagulation.

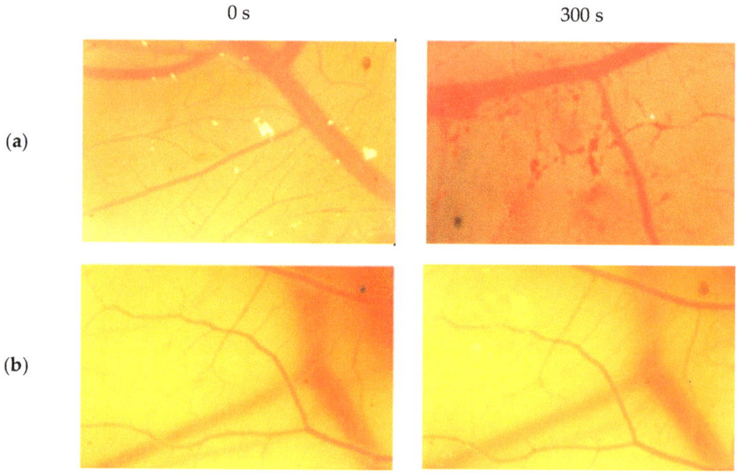

Figure 4. *Cont.*

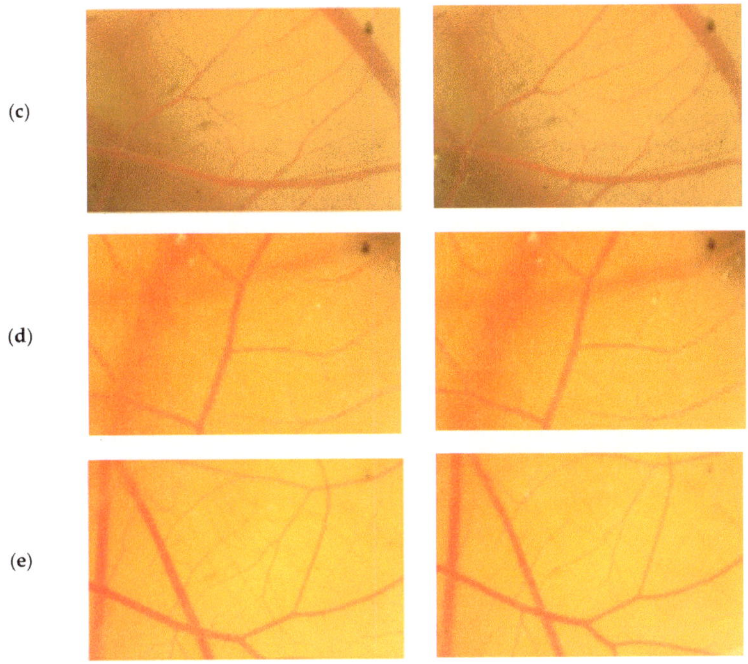

Figure 4. Vessels of sodium dodecyl sulfate (SDS) (**a**), sample A (**b**), sample B (**c**), sample C (**d**), and NaCl (**e**) from 0 s and 300 s.

4. Discussion

PDMS has been successfully synthesized using a low-grade D4 monomer with different synthesis conditions from PDMS synthesized using a high-grade D4 monomer. Sample A synthesized with higher synthesis temperature and KOH concentration compared to sample B. In addition, the time of polymerization of sample A is also longer than sample B. However, simply by changing the volume ratio of D4 and MM used from 26:10 to 46:10, medium viscosity of PDMS can be produced from the same purity monomer.

Using a low grade of D4 monomer gives some excellent properties, especially for surface tension and yield value of the samples. The high surface tension of sample A and sample C will reduce the possibility of emulsification. However, these samples are still easy to use in vitreoretinal surgery due to their viscosity. The yield value of the samples will affect the number of products produced. The higher the yield values of the samples, the better the production effectiveness. Samples from a low grade of D4 monomer have a higher refractive index than the sample from a high grade of D4 monomer. However, the additional diopters are still in the range of allowable values (+3.0D until +3.5D) [15]. Moreover, sample A and sample C have the appearance as transparent as sample B, as shown in Figure 1.

The IR spectra of all samples showed that all the samples had similar transmittance peaks to commercial but with different intensities. The sample with higher viscosity has lower peak intensity. However, the sample with similar viscosity using a low-grade D4 monomer had higher intensity than sample using a high-grade D4 monomer. The main functional groups of PDMS were found in all samples without any impurities. The FTIR result confirmed that all the samples were PDMS.

The low grade of D4 monomer has successfully produced the good quality of low viscosity and medium viscosity of PDMS. However, if the safety of PDMS is not guaranteed, it will be nothing. For this reason, the toxicological test must be carried out before it is used in the human eyes. Chorioallantoic Membrane (CAM) is a complete tissue in the hen's eggs and consists of arteries, veins, and capillaries. CAM has a similar reaction process as in rabbit eyes (Draize test), especially in conjunctiva tissue, when it is exposed to an irritating substance. Hen embryos have been used in various fields of medical research since the 20th century. The egg that was used in this test is the fertile white egg. The egg was chosen based on the vascular condition during candling CAM process [11]. Based on interlaboratory validity, which was conducted by Hagino in 1999, the HET-CAM method can provide an alternative method to evaluate potential irritation of chemicals to the eye. CAM test evaluates vascular reactions and damage of the CAM in the presence of macroscopic changes such as hemorrhage, lysis, and coagulation. Previous studies showed that HET-CAM and in vivo eye irritation tests have a good correlation [16].

Compared with other alternative tests, the HET-CAM test has some good points, such as could be carried out for all types of chemical substance (liquid or powder), applied with the similar condition of in vivo test, easy to access for research, not expensive, not need complicated animal room facility, and simple method [10,16]. However, the HET-CAM test is not suitable for red-pigmented samples. To maintain objectivity in the evaluation process of potential eye irritation, the HET-CAM test requires the use of a reference substance. In addition, the HET-CAM test also needs an experienced investigator [16].

Evaluation of potential irritants of the test substance is spelled out with the scoring system. A scoring system has gradually been developed by selecting various criteria to obtain the most consistent correlation with the variation in the concentration of the preparation being tested.

In vitro toxicity test showed that all types of damage occur in positive control. Hemorrhage occurred in all groups starting at 10 s. Lysis occurred from the 30 s. Meanwhile, coagulation occurred from the 60 s. A hemorrhage is a condition when blood comes out of damaged blood vessels. Lysis is when the integrity of the cell membrane is broken or damaged and causes cell organelle to come out. Coagulation is a condition when the blood freezes. Samples were non-irritant through the HET-CAM test. Based on HET-CAM test evaluation, negative control (NaCl 0.9%) was non-irritant, positive control (1% SDS) was a strong irritant, and all PDMS samples get zero (0) score that indicated as non-irritant substances.

Several studies have shown that the low viscosity of PDMS has a higher emulsification tendency than other viscosities. This is influenced by low surface tension and impurities [3]. Both cause ocular toxicity by spreading to the eye tissue. In this study, we successfully repeated the synthesis of low and medium viscosity PDMS with high surface tension. In vitro toxicity tests showed that low-viscosity PDMS from low-grade D4 monomers did not cause irritation, as did PDMS synthesized from high-grade D4 monomers. The results of the PDMS toxicity test with medium viscosity also showed that this PDMS is a non-irritating substance. These results prove that samples with high surface tension values have low emulsification tendencies and reduce sample toxicity. In addition, the use of low-grade D4 monomers can produce high-quality PDMS as well as PDMS synthesized using high-grade D4 monomers.

The result of in vitro toxicity test by the HET-CAM test method is appropriate with the other in vitro toxicity test results. Romano et al. explained that the cytotoxic effect of low viscosity of PDMS, with different purification levels, was not found in human retinal cells (ARPE-19 and BALB 3T3) [5]. Moreover, these in vitro toxicity test results are also suitable within in vivo toxicity tests in rabbit eyes. Mackiewicz et al. explained that low viscosity (1000 mPa.s) and medium viscosity (3000 mPa.s) of PDMS show no signs of inflammation or hyperemia after 3 months tested into rabbit eyes [17].

5. Conclusions

We have successfully synthesized low- and medium-viscosity PDMS materials using a low-grade D4 monomer and characterized the physical properties, including the in vitro toxicity test by the HET-CAM test method. The characterization of physical properties showed that the refractive index, surface tension, and functional groups of all PDMS samples were very similar to commercial PDMS and PDMS synthesized using a high-grade D4 monomer. The addition of diopters of PDMS samples is also within the allowed normal range. The results of the in vitro toxicity test using the HET-CAM method showed that PDMS with low and medium viscosity was a non-irritant substance indicating that all samples were non-toxic. These results proved that the use of low-grade D4 monomer can produce high-quality PDMS, is safe, and has the potential to be developed as a substitute for vitreous humor in the human eye.

Supplementary Materials: The following supporting information can be downloaded at: https://www.mdpi.com/article/10.3390/jfb13010003/s1, Figure S1. Hemorrhage score of (a) egg 1, (b) egg 2, and (c) egg 3, Figure S2. Lysis score of (a) egg 1, (b) egg 2 and (c) egg 3, Figure S3. Coagulation score of (a) egg 1, (b) egg 2 and (c) egg 3, Figure S4. Vessels of positive control, Figure S5. Vessels of sample A, Figure S6. Vessels of sample B, Figure S7. Vessels of sample C. Figure S8. Vessels of the negative control.

Author Contributions: Investigation, formal analysis, writing—original draft preparation, D.G.A.; investigation, data curation, S.S.; conceptualization, methodology, F.F.; conceptualization, validation, writing—review and editing, supervision R.R. All authors have read and agreed to the published version of the manuscript.

Funding: This research was funded by Kemenristek of Indonesia in the scheme of Fundamental Research (Penelitian Dasar) 2021, contract No. 1207/UN6.3.1/PT.00/2021, and also partially sup-ported by Academic Leadership Grant of Universitas Padjadjaran 2021, contract No. 1959/UN6.3.1/PT.00/2021.

Institutional Review Board Statement: Not applicable.

Informed Consent Statement: Not applicable.

Data Availability Statement: Not applicable.

Acknowledgments: Authors would like to thank Kemenristek of Indonesia for financial support in the scheme of Fundamental Research (Penelitian Dasar) 2021, contract No. 1207/UN6.3.1/PT.00/2021. This work was also partially supported by the Academic Leadership Grant of Universitas Padjadjaran 2021, contract No. 1959/UN6.3.1/PT.00/2021.

Conflicts of Interest: The authors declare no conflict of interest.

References

1. Kartasasmita, A.; Kusdiono, W.; Virgana, R.; Boesorie, S. In Vivo Emulsification Analysis of 1000 cs and 5000 cs Silicone Oil after Rhegmatogenous Retinal Detachment Vitrectomy Surgery. *Open J. Ophthalmol.* **2007**, *7*, 231–239. [CrossRef]
2. Giordano, G.G.; Refojo, M.F. Silicone Oils as Vitreous Substitutes. *Prog. Polym. Sci.* **1998**, *23*, 509–532. [CrossRef]
3. Nusa, H.S.; Astuti, W.; Kartasasmita, A.S.; Virgana, R.; Syakir, N.; Bahtiar, A.; Safriani, L.; Risdiana. Characterization of Optical and Structure Properties of Polydimethylsiloxanes. *Mater. Sci. Forum* **2015**, *827*, 99–104. [CrossRef]
4. Caramoy, A.; Schröder, S.; Fauser, S.; Kirchhof, B. In Vitro Emulsification Assessment of New Silicone Oils. *Br. J. Ophthalmol.* **2010**, *94*, 509–512. [CrossRef] [PubMed]
5. Romano, M.; Ferrara, M.; Gatto, C.; Giurgola, L.; Zanoni, M.; Angi, M.; Rinaldi, M.; Borgia, A.; Sorrentino, T.; Tothova, J.D. Safety of Silicone Oils as Introcular Medical Device: An In Vitro Cytotoxicity Study. *Exp. Eye Res.* **2020**, *194*, 108018. [CrossRef] [PubMed]
6. Nakamura, K.; Refojo, M.F.; Crabtree, D.V.; Pastor, J.; Leong, F.L. Ocular Toxicity of Low-Molecular-Weight Components of Silicone and Fluorosilicone Oils. *Investig. Ophthalmol. Vis. Sci.* **1991**, *32*, 3007–3020.
7. Brunner, S.; Izay, B.; Weidinger, B.; Maichel, B.; Binder, S. Chemical Impurities and Contaminants in Different Silicone Oils in Human Eyes before and After Prolonged Use. *Graefe's Arch. Clin. Exp. Ophthalmol.* **2010**, *249*, 29–36. [CrossRef] [PubMed]
8. Tavaszi, J.; Budai, P. Toxicity study of agrochemicals on chorioallantoic membrane of the egg. *Commun. Agric. Appl. Biol. Sci.* **2006**, *71*, 101–105. [PubMed]
9. Walum, E.; Balls, M.; Bianchi, V.; Blaauboer, B.; Bolcsfoldi, G.; Guillouzo, A.; Moore, G.A.; Odland, L.; Reinhardt, C.; Spielmann, H. ECITTS: An integrated approach to the application of in vitro test systems to the hazard assessment of chemicals. *Altern. Lab. Anim.* **1992**, *20*, 406–428. [CrossRef]
10. Leighton, J.; Nassauer, J.; Tchao, R. The Chick Embryo in Toxicology: An Alternative to the Rabit Eye. *Food Chem. Toxicol.* **1985**, *23*, 293–298. [CrossRef]

11. Valdes, T.I.; Kreutzer, D.; Moussy, F. The chick chorioallantoic membrane as a novel in vivo model for the testing of biomaterials. *J. Biomed. Mater. Res.* **2002**, *82*, 273–282. [CrossRef] [PubMed]
12. Auliya, D.G.; Setiadji, S.; Agasa, Z.M.; Fitrilawati; Syakir, N.; Risdiana. Synthesis of Low Viscosity Polydimethylsiloxane Using Low Grade of Octamethylcyclotetrasiloxane. *Mater. Sci. Forum* **2021**, *1028*, 365–370. [CrossRef]
13. Setiadji, S.; Agasa, Z.M.; Auliya, D.G.; Fitrilawati; Syakir, N.; Noviyanti, A.R.; Rahayu, I.; Supriadin, A.; Risdiana. Synthesis and Characterization of Polydimethylsiloxane (PDMS) with Medium Viscosity via Ring-Opening Polymerization. *Mater. Sci. Forum* **2021**, *1028*, 346–351. [CrossRef]
14. ICCVAM. Recommended Test Method Protocol: Hen's Egg Test–Chorioallantoic Membrane (HET-CAM) Test Method. Available online: http://iccvam.niehs.nih.gov/methods/ocutox/MildMod-TMER.htm (accessed on 28 April 2021).
15. Swindle, K.E.; Ravi, N. Recent advances in polymeric vitreous substitutes. *Expert Rev. Ophthalmol.* **2007**, *2*, 255–265. [CrossRef]
16. Hagino, S.; Kinoshita, S.; Tani, N.; Nakamura, T.; Ono, N.; Konishi, K.; Iimura, H.; Kojima, H.; Ohno, Y. Interlaboratory validation of in vitro eye irritation tests for cosmetic ingredients. (2) Chorioallantoic membrane (CAM) test. *Toxicol. Vitro* **1999**, *13*, 99–113. [CrossRef]
17. Mackiewicz, J.; Muhling, B.; Hiebl, W.; Meinert, H.; Maaijwee, K.; Kociok, N.; Luke, C.; Zagorski, Z.; Kirchhof, B.; Joussen, A.M. In Vitro Retinal Tolerance of Various Heavy Silicone Oils. *Investig. Ophthalmol. Vis. Sci.* **2007**, *48*, 1873–1883. [CrossRef] [PubMed]

Journal of
*Functional
Biomaterials*

MDPI

Article

Effects of Silver Diamine Nitrate and Silver Diamine Fluoride on Dentin Remineralization and Cytotoxicity to Dental Pulp Cells: An In Vitro Study

Surapong Srisomboon [1], Matana Kettratad [2], Andrew Stray [3], Phakkhananan Pakawanit [4], Catleya Rojviriya [4], Somying Patntirapong [2,5] and Piyaphong Panpisut [2,5,*]

1 Department of Oral Health Care, Lamlukka Hospital, Pathum Thani 12150, Thailand; surapong.sri@dome.tu.ac.th
2 Faculty of Dentistry, Thammasat University, Pathum Thani 12120, Thailand; pmatana@staff.tu.ac.th (M.K.); p_somying@hotmail.com (S.P.)
3 DentaLife, Ringwood, VIC 3134, Australia; astray@dentalife.com.au
4 Synchrotron Light Research Institute (Public Organization), Nakhon Ratchasima 30000, Thailand; phakkhananan@slri.or.th (P.P.); catleya@slri.or.th (C.R.)
5 Thammasat University Research Unit in Dental and Bone Substitute Biomaterials, Thammasat University, Pathum Thani 12120, Thailand
* Correspondence: panpisut@tu.ac.th

Abstract: Silver diamine nitrate (SDN) is expected to help control caries similar to silver diamine fluoride (SDF). The aim of this study was to determine the mineral precipitation in demineralized dentin and the cytotoxicity of SDN and SDF to dental pulp cells. Demineralized dentin specimens were prepared, and SDF, SDN, or water (control) was applied. The specimens were then remineralized in simulated body fluid for 2 weeks. The mineral precipitation in the specimens was examined using FTIR-ATR, SEM-EDX, and synchrotron radiation X-ray tomographic microscopy (SRXTM). Additionally, the cytotoxicity of SDF and SDN to human dental pulp stem cells was analyzed using an MTT assay. The increase in FTIR spectra attributable to apatite formation in demineralized dentin in the SDF group was significantly higher compared to the SDN and control groups ($p < 0.05$). Dentinal tubule occlusion by the precipitation of silver salts was detected in both SDF and SDN groups. The mineral density as shown in SRXTM images and cytotoxicity of both SDN and SDF groups were comparable ($p > 0.05$). In conclusion, SDF demonstrated superior in vitro apatite formation compared to SDN. However, the degree of mineral precipitation and cytotoxic effects of both were similar.

Keywords: silver diamine nitrate; silver diamine fluoride; dental caries; tooth demineralization; tooth remineralization; cytotoxicity test; dental pulp; synchrotron; X-ray microtomography

1. Introduction

Untreated dental caries represent the most common preventable chronic disease affecting people of all ages worldwide [1]. A study showed that at least 1 in 5 adults in the U.S. population have untreated caries [2]. Current cost-effective cavity management consists of delaying irreversible surgical treatment and promoting remineralization to arrest the progression of lesions [3]. Low-invasive methods are also suitable for patients with special needs or with limited cooperation. The most common non-invasive materials for controlling dental caries are professionally applied fluoride materials such as silver diamine fluoride (SDF) [4] and NaF varnish [5]. The use of SDF is a cost-effective method of arresting dental caries [6,7]. It was demonstrated that biannual application of SDF led to a higher level of prevention of caries progression than NaF varnish [8].

There are four main anti-caries effects from SDF. The first is direct antibacterial action from the silver ion of SDF [9]. The second effect is the precipitation of silver phosphate or silver chloride [10], which can potentially enhance lesion hardness and act as a protective layer against dental biofilm. The third action is the formation of low-soluble and acid-resistant fluorohydroxyapatite, which can increase resistance to caries for the tooth surface. The fourth action is the ability to preserve collagen in dentin, which is essential for mineral precipitation. Silver ion was shown to reduce the degradation of collagen, which acts as a template for mineral precipitation [11]. The SDF solution can be rapidly adsorbed into dentin. The concern was that the high level of reactive ions in SDF could induce cytotoxic effects on the dentin–pulp complex. It was demonstrated that reactive ions such as Ag ion could diffuse into dentin up to 5–40 μm [12]. It was reported that the toxic effects of Ag and F ions are depleted glutathione and increased oxidative stress or lipid peroxidation [13]. This leads to reduced antioxidant properties, resulting in cell death and inflammation. Additionally, it was reported that the cytotoxic action of hydroxyapatite disc treated with SDF persisted even after 77 days of water rinsing [14]. A study also proposed applying glutathione with SDF to promote antioxidant functions and decrease the toxic effects from SDF on dental pulp cells [13].

An alternative silver solution for controlling caries is silver nitrate ($AgNO_3$). An in vitro study showed that the application of $AgNO_3$ or AgF increased the mineral density of demineralized enamel and dentin [15]. The increased mineralization was believed to be mainly due to the deposition of silver ions. Additionally, many studies have demonstrated that the use $AgNO_3$ in combination with NaF varnish had comparable effectiveness in caries prevention compared to SDF [16–19]. This method may be more feasible, with a lower cost, compared to SDF, since $AgNO_3$ solution and NaF varnish are already available in many countries [20]. However, the concern with $AgNO_3$ is the delay in mineral induction time, which may affect remineralization in dentin [21]. SDF contains diamine groups, which may enable the formation of NH_4OH, which could potentially help promote suitable pH and conditions for mineral formation and enhance antibacterial action [22,23]. The addition of diamine groups to stabilize silver ions in $AgNO_3$ nitrate solution, forming silver diamine nitrate (SDN), is expected to help enhance the mineral precipitation of the solution.

Although SDF is considered a cost-effective intervention for controlling caries, the cost of the materials can vary across regions. Additionally, SDF may still not be available in some countries [24]. The cost of SDF in the U.S. was approximately USD 30–52 per application [6]. From the manufacturer's point of view, the cost of an alternative silver solution such as silver diamine nitrate (SDN) is expected to be lower than that of SDF due to the lack of fluoride components. This would help reduce the economic burden for cavity prevention programs [25]. Currently, in vitro assessment of the remineralizing effects of SDN is limited. The aim of the current study was, therefore, to compare the mineral precipitation in demineralized dentin and the cytotoxicity of pulp cells between silver diamine nitrate (SDN) and silver diamine fluoride (SDF). It was expected that SDN would encourage mineral precipitation similar to SDF, and that the cytotoxic effects of SDN on dental pulp cells would be comparable with those of SDF. The null hypothesis was that the in vitro mineral precipitation and relative cell viability between SDF and SDN would not be significantly different.

2. Materials and Methods

2.1. Specimen Preparation for Remineralizing Studies

Extracted human third permanent molars of comparable size and with no visible cavitated carious lesions were collected from the Department of Oral Health Care, Thammasat University Hospital, Pathum Thani, Thailand. The use of human teeth was approved by the Ethics Review Sub-Committee for Research Involving Human Research Subjects at Thammasat University (approval number: 150/2562). The teeth were stored for less than 30 days in 0.1% thymol solution (M-Dent, Faculty of Dentistry, Mahidol University, Bangkok, Thailand) at 23 °C prior to the experiment.

Specimen preparation was performed according to the protocol used in the previous study [26]. Briefly, the teeth were embedded in self-curing acrylic resin ($n = 7$). The crown of each tooth was sectioned horizontally and perpendicular to dentinal tubules. The diamond blade of the cutting machine (Accutom 50, Struers, Cleveland, OH, USA) was positioned at ~2 mm below the occlusal surface. The obtained dentin slices (2.0 $\pm$ 0.1 mm thick) were then polished with microfine 4000-grit abrasive paper in a polishing machine (Tegramin, Struers, Cleveland, OH, USA). Then, the specimens were cleaned in an ultrasonic bath for 5 min. Each dentin slice was cut into 3 pieces using a greater taper medium-fine diamond bur to produce a total of 21 dentin specimens.

The specimens were demineralized in 17% ethylenediamine tetraacetic acid (EDTA; Faculty of Dentistry, Chulalongkorn University, Bangkok, Thailand) for 72 h to produce completely demineralized layers (depth of ~500 μm) [27,28]. Then, 25 μL of silver diamine nitrate (SDN) solution (48% SDN; Dentalife, Victoria, Australia), SDF (38% SDF, Topamine^{TM}; Dentalife, Victoria, Australia), or deionized water (control group) was applied to specimens from each tooth (n = 7/group) for 30 s. The specimens were then cleaned with water from a triple syringe for 10 s, and immersed in simulated body fluid (SBF; BS ISO 23317:2014) (Table 1) [29]. SBF contains the same phosphate concentration as blood plasma or body fluid (pH = 7.40) (Table 2) [30,31]. SBF was expected to mimic the environment where the solution was adsorbed into the dentin and exposed to dentinal fluid. The specimens were incubated at 37 °C for up to 2 weeks without replacing the solution.

Table 1. Chemicals used to prepare SBF in the current study. All chemicals were purchased from Sigma Aldrich (St. Louis, MO, USA).

Order	Chemical	Amount (g)
1	NaCl	8.035
2	$NaHCO_3$	0.355
3	KCl	0.225
4	$K_2HPO_4 \cdot 3H_2O$	0.231
5	$MgCl_2 \cdot 6H_2O$	0.311
6	HCl (1 M)	38
7	$CaCl_2 \cdot 2H_2O$	0.386
8	Na_2SO_4	0.072
9	Tris, $NH_2C(CH_2OH)_3$	6.118

Table 2. Concentration (10^{-3} mol) of ions in SBF and blood plasma.

Ion	SBF (pH 7.4)	Blood Plasma (pH 7.2–7.4)
Na^+	142.0	142.0
K^+	5.0	5.0
Mg^{2+}	1.5	1.5
Ca^{2+}	2.5	2.5
Cl^-	147.8	103.0
HCO_3^-	4.2	27.0
HPO_4^{2-}	1.0	1.0
SO_4^{2-}	0.5	0.5

2.2. Assessment of Apatite Precipitation Using FTIR and SEM-EDX

Apatite formation on the demineralized dentin was examined using a Fourier transform infrared spectrometer equipped with attenuated total reflection (FTIR-ATR; Nicolet iS5, Thermo Fisher Scientific, Waltham, MA, USA) (n = 7) [26,32–35]. FTIR spectra in the region of 700–4000 cm^{-1} (resolution of 8 cm^{-1} with 12 repetitions) were recorded from the bottom surface of the specimen. The FTIR spectra of specimens were recorded after demineralization, then after remineralizing in SBF for 1 day, 1 week, and 2 weeks.

The ratio of FTIR area attributed to hydroxyapatite (1024 cm^{-1}, PO_4^{3-} stretch) [36] over the peak representing type I collagen in dentin (1636 cm^{-1}, C=O stretch of amide I) [37] was obtained using OMNIC Series software (Thermo Fisher Scientific, Waltham, MA, USA). The mineral/matrix ratio (Abs_{1024}/Abs_{1636}) was then calculated. An increase in the Abs_{1024}/Abs_{1636} ratio was expected to relate to an increase in mineral precipitation (remineralization) in demineralized dentin [26].

A representative specimen from each group was then selected to assess the mineral precipitation on the surface. The specimens were coated with gold in a sputter-coating machine (Q150R ES, Quorum Technologies, East Sussex, UK) using a 23 mA current for 45 s. A dispersive X-ray spectrometer (EDX, X-Max 20, Oxford Instruments, Abingdon, UK) was employed to analyze the elemental composition of precipitation on the specimens. The EDX spectrum was obtained from the precipitate using magnification of 20,000× and beam voltage of 5 kV. Data were then analyzed using INCA software version 5.05 (ETS, Stuttgart, Germany).

2.3. Assessment of Mineral Precipitation Using Synchrotron-Based X-ray Tomography (SRXTM)

Representative specimens (n = 3) at 2 weeks were selected and blotted dry (n = 3). The mineral density in the demineralized area was examined by a synchrotron X-ray source, Beamline 1.2 W X-ray imaging and tomographic microscopy (XTM), according to the method used in a previous study [26]. The synchrotron X-ray radiation originated from a 2.2-Tesla multipole wiggler at the Siam Photon Source operated at 1.2 GV. By using a polychromatic X-ray beam with a distance from source to sample of 32 m, the experiments were executed at a mean energy of 14 kV. Representative specimens were mounted on the stage. Then, X-ray radiographs were collected from 0° to 180° with an angular increment of 0.2°. The collected X-ray radiographs were then analyzed using Octopus Reconstruction software (TESCAN, Gent, Belgium) [38] to produce reconstruction images. After obtaining the reconstruction images, the degree of mineral precipitation was calculated by using Octopus Analysis software. In this case, 200 reconstruction images (or 288 μm) were chosen and averaged by 3 random areas (~10 × 10 μm). The reconstruction images were computed by using Drishti software [39] to produce the 3D tomographic reconstruction.

2.4. Cytotoxicity Test

Human dental pulp stem cells (hDPSCs) were obtained from Lonza (PT-5025, Group AG, Basel, Switzerland). Cells were maintained in Dental Pulp Stem Cell Basal Medium supplemented with Dental Pulp Stem Cell Growth Supplement, L-glutamine, ascorbic acid, and gentamycin/amphotericin-B (all from Lonza) at 37 °C enriched with 5% CO_2. For the experiment, hDPSCs at passage 3 were switched to culture in Dulbecco's Modified Eagle Medium (Sigma-Aldrich, St. Louis, MO, USA) with 10% fetal bovine serum and 1% penicillin/streptomycin and seeded in a 96-well plate with a cell density of 5,000 cells/well. The cells were then treated with 25 μL of SDF or SDN. Cells with no treatment were used as the control. The cells were cultured at 37 °C enriched with 5% CO_2 for 3 days. Then, an MTT viability assay was performed. DPSCs were incubated with 0.2% 3-(4,5 dimethylthiazolyl)-2,5-di-phenyltetrazolium bromide (MTT) solution (Sigma-Aldrich, St. Louis, MO, USA) at 37 °C for 4 h. The reaction was paused using 200 μL of dimethylsulfoxide (Sigma-Aldrich, St. Louis, MO, USA) and 25 μL glycine buffer (Research Organics, Cleveland, OH, USA). The color of the end product was quantified using absorbance at 620 nm [40,41] under a spectrophotometer (Sunrise Absorbance Microplate Reader, Tecan Group Ltd., Männedorf, Switzerland). The results were expressed as relative optical density (OD) at 620 nm using the following equation:

$$\text{Relative OD} = \frac{\text{OD of test group}}{\text{OD of control}} \times 100 \tag{1}$$

2.5. Statistical Analysis

Data were analyzed using Prism 9 for macOS (GraphPad Software, San Diego, CA, USA). The normality of data was initially examined using the Shapiro–Wilk test. Changes in Abs_{1024}/Abs_{1636} for the same group upon immersion time were compared using one-way repeated ANOVA followed by Tukey's multiple comparisons. Differences in Abs_{1024}/Abs_{1636} and mineral density were determined using one-way ANOVA followed by Tukey's multiple comparisons. Additionally, the unpaired t-test was used to compare relative OD values between SDF and SDN groups. All p-values below 0.05 are considered statistically significant. A post hoc power analysis was performed using G*Power version 3.1.9.6 (University of Dusseldorf, Dusseldorf, Germany). The effect size [42] of each experiment was calculated from the results obtained from the previous study [26], demonstrating that the sample size used in each test exhibited power > 0.95 at alpha = 0.05.

3. Results

3.1. Assessment of Apatite Precipitation Using FTIR and SEM-EDX

The FTIR spectra of the representative specimens at each time point are presented in Figure 1. A reduced phosphate peak (1024 cm^{-1}) after demineralization was observed in all groups. The Abs_{1024}/Abs_{1636} ratio of the control group at 0 h (0.22 $\pm$ 0.06) was significantly higher than that of specimens at 24 h (0.11 $\pm$ 0.04) (p = 0.0285), 168 h (0.10 $\pm$ 0.03) (p = 0.0148), and 336 h (0.11 $\pm$ 0.03) (p = 0.0377) (Figure 2). The mean Abs_{1024}/Abs_{1636} ratio of the SDF group increased from 0.20 $\pm$ 0.10 at 0 h to 0.30 $\pm$ 0.16 at 336 h. The Abs_{1024}/Abs_{1636} ratio of the SDN group at 0 h was 0.32 $\pm$ 0.15, which was gradually reduced to 0.25 $\pm$ 0.22 at 336 h. However, changes in the Abs_{1024}/Abs_{1636} ratio for the SDF and SND groups at each time point were not significantly different ($p > 0.05$). Additionally, the ratio between SDF, SDN, and control groups at each time point was comparable ($p > 0.05$).

The difference in Abs_{1024}/Abs_{1636} of the SDF group at 336 h compared with 0 h (0.10 $\pm$ 0.09) was significantly higher than that of the SDN (−0.07 $\pm$ 0.12) (p = 0.0140) and control (−0.11 $\pm$ 0.08) (p = 0.0026) groups (Figure 3). Additionally, the difference in the ratio was not significantly different between the SDN and control groups (p = 0.7175).

SEM images of the representative specimen of the control group show patent dentinal tubules (Figure 4A). SEM images of the SDN (Figure 4B) and SDF (Figure 4C) groups show crystals occluding dentinal tubules. The EDX results indicate that the precipitation observed in the SDN and SDF groups mainly contained Ag and Cl (Figure 4D).

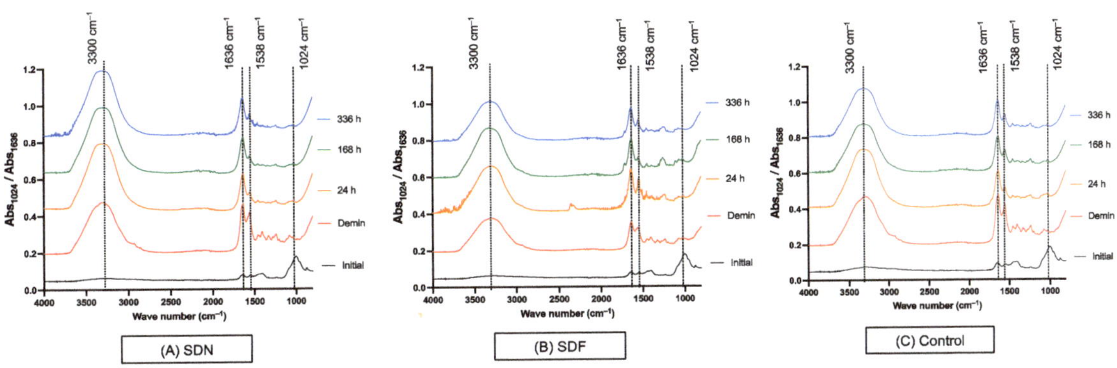

Figure 1. FTIR spectra of representative specimen from (**A**) SDN, (**B**) SDF, and (**C**) control groups at each time point.

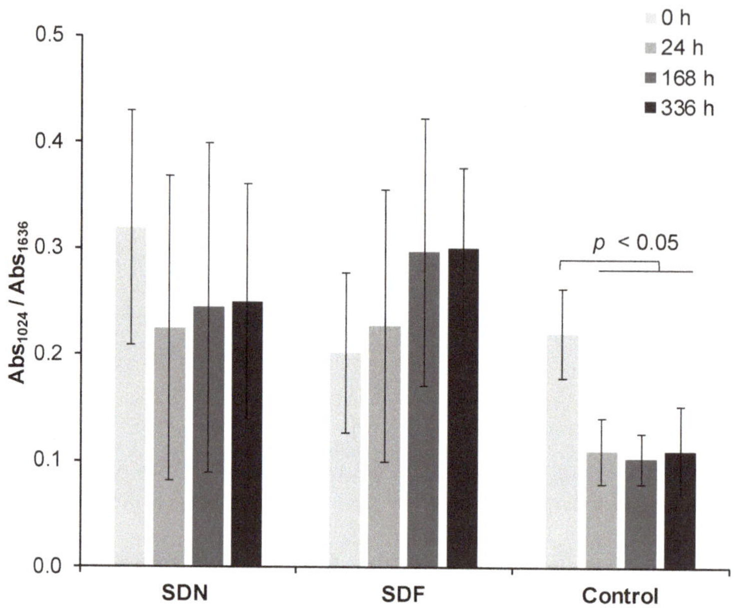

Figure 2. Abs_{1024}/Abs_{1636} ratio of demineralized dentin specimens before and after immersing in simulated body fluid for 336 h (2 weeks). Error bars represent SD ($n = 7$), lines indicate p-values.

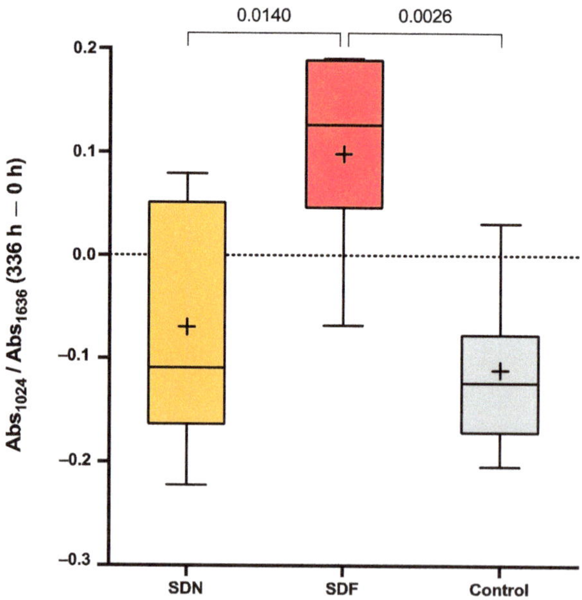

Figure 3. Box plots of differences in Abs_{1024}/Abs_{1636} after 2 weeks compared with initial value (336 vs. 0 h). Boxes represent first quartile (Q1) to third quartile (Q3), horizontal lines in box represent median, and whiskers represent maximum and minimum values ($n = 7$). Lines indicate $p < 0.05$.

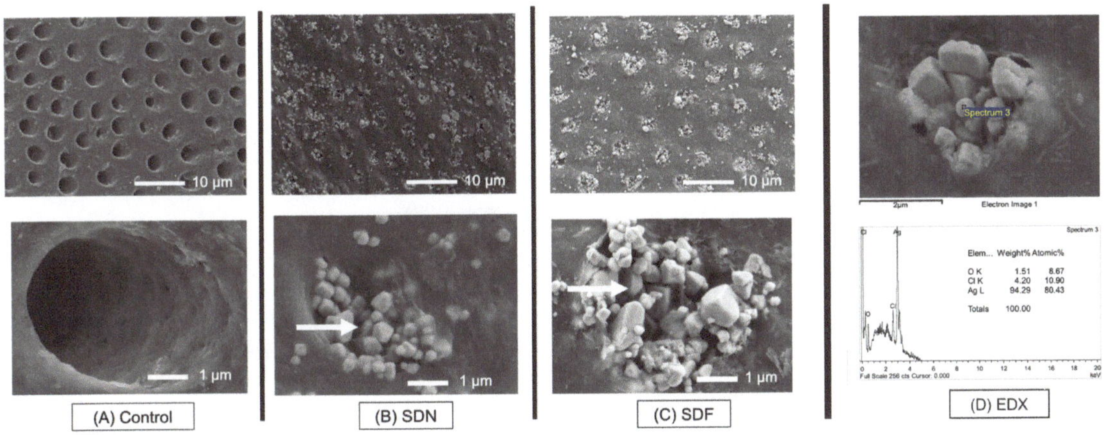

Figure 4. (**A–C**) SEM images at low and high magnification of representative specimen from each group after 2 weeks. Precipitation of crystals (arrows) occluding dentinal tubules was observed in SDN and SDF groups. (**D**) EDX result shows that crystals mainly contained Ag and Cl.

3.2. Assessment of Mineral Precipitation Using Synchrotron-Based X-ray Tomography (SRXTM)

The SRXTM images of representative specimens of the SDN (Figure 5A, Video S1) and SDF (Figure 5B, Video S2) groups show multiple radiodense areas throughout the depth of the radiolucent area (~200 µm). No radiodense areas were detected in the specimen from the control group (Figure 5C, Video S3).

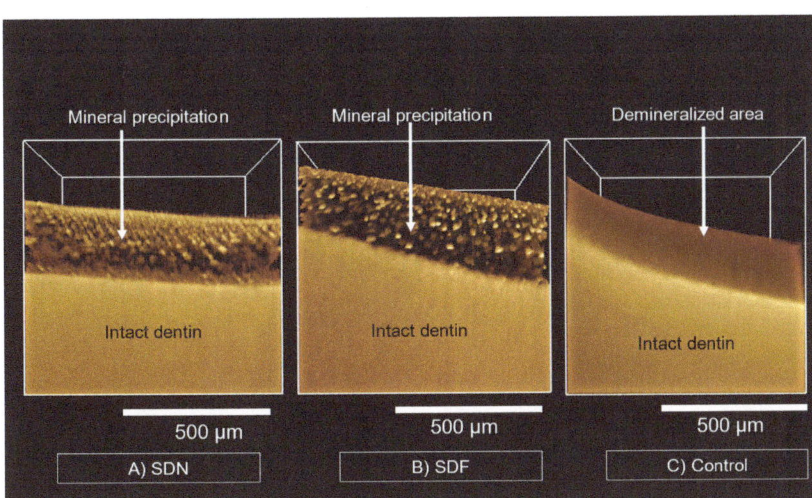

Figure 5. SRXTM images of representative specimen from each group after immersion in simulated body fluid for 2 weeks. (**A**,**B**) Mineral precipitation (arrows) was detected in SDN and SDF groups. (**C**) No precipitation was detected in control group. Three-dimensional images of specimens are provided in the Supplementary Materials (Videos S1–S3).

3.3. Cell Viability

The relative OD of the SDF group (92 ± 8%) was higher than that of the SDN group (88 ± 10%). However, the results were not significantly different (p = 0.5579).

4. Discussion

The aim of the current study was to compare the remineralizing and cytotoxic effects of silver diamine nitrate (SDN) and silver diamine fluoride (SDF). The increased Abs_{1024}/Abs_{1636} ratio of the SDF group was significantly higher than that of the SDN and control groups. However, the mineral precipitation, as seen in SRXTM images, and the cytotoxicity of the SDN and SDF groups were comparable. Hence, the null hypothesis of the current study was partially rejected. It should be mentioned that study was an in vitro study; thus, the clinical relevance should be interpreted with caution.

The FTIR-ATR results of the current study indicate that the increase in peaks representing apatite formation was greater in the SDF group than the SDN group. This could be due to the effects of fluoride. It was demonstrated that fluoride can act as a catalyst for phosphate and calcium ion addition into the crystal lattice, thus promoting the growth of apatite crystals [43]. This may have subsequently promoted mineral apatite formation in the SDF specimen. Additionally, it was reported that the incorporation of fluoride into the lattice of remnant crystals promoted crystal transformation and reduced the solubility of apatite [44]. This study aimed to assess mineralization when the solution penetrates into the dentin and is exposed to dentinal fluid. However, it should be mentioned that a protocol for preparing simulated dentinal fluid has not yet been established. We speculated that the inorganic components of dentinal fluid would be similar to other types of body fluid [45]. Simulated body fluid (SBF) was therefore selected as the storage solution in the current study.

A high level of fluoride from SDF was expected to encourage the formation of low-soluble fluorohydroxyapatite (FHA) in the demineralized dentin. It is known that the precipitation of FHA after SDF application is not easily detected. This is in agreement with the current study, because the EDX failed to detect fluoride on the specimens. This could be due to the low sensitivity of EDX to detect small amounts of fluoride. Hence, the use of an alternative technique such as XRD, XPS, SAX/WAX, or Raman microcopy may be needed in future work to confirm the formation of fluorohydroxyapatite. Additionally, the application of SDF may encourage the precipitation of CaF_2, which can act as a fluoride reservoir [46,47]. However, CaF_2 globules were not detected in the SEM images of specimens. This could be due to the rapid washout of water-soluble CaF_2 during rinsing with water [47,48].

The SEM images showed substantial mineral crystals occluding the patent dentinal tubules on the surface of specimens from the SDF and SDN groups. EDX showed that the crystals mainly contained Ag and Cl, indicating the formation of silver chloride salts. It is believed that Ag ions of SDF and SDN react with ions in the environment, producing silver phosphate (solubility of 6.4×10^{-3} g/100 mL) and silver oxide (1.3×10^{-3} g/100 mL), which are highly soluble. Then, the silver compounds may readily react with chloride in the environment, forming lower-soluble silver chloride (8.9×10^{-5} g/100 mL) [47,49]. The formation of silver chloride caused a black/metallic appearance on the tooth surface after SDF application. The formation of silver salts increased the hardness of dentin and blocked dentinal tubules, thus reducing the irritation on the pulp–dentin complex [9,47].

The use of synchrotron radiation X-ray tomographic microscopy (SRXTM) enabled 3D visualization of mineral precipitation at a higher resolution (pixel size ~1.44 μm) compared with conventional micro-CT (pixel size ~8 μm) [50]. The more radiodense areas in SRXTM images were expected to be silver salts [26]. The degree of mineral precipitation was similar in the SDN and SDF groups. This may be due to a comparable concentration of silver ions contained in both (25 ppm). However, the limitation of SRXTM is the risk of overestimating mineral density due to the high radiopacity of silver [51]. Additionally, the density of mineral crystals per volume may be affected by the dehydration and shrinkage of demineralized dentin layers. It should be mentioned that particles with a diameter smaller than the minimum resolution of the SRXTM (1.44 μm) were not detected in the images. This may lead to underestimation of the mineral density of small nanosized crystals such as hydroxyapatite in specimens [52].

The SDF solution can be rapidly adsorbed into dentin. Thus, the concern with applying SDF in deep cavities is the risk of toxic effects on dental pulp cells, leading to pulpal pain and inflammation. It has been shown that SDF reduces pulpal-like cell viability via the depletion of glutathione [13]. The use of high-molecular weight nitrate molecules in the solution was expected to lower the reactivity of the reactive ions, which could potentially reduce the toxic effects on pulp cells. However, the relative OD values of SDF and SDN groups were comparable. This could be due to the similar concentrations of silver ions in SDF and SDN in the current study. BS EN ISO 10993–5: 2009 (Biological evaluation of medical devices Part 5: Tests for in vitro cytotoxicity) states that a reduction in cell viability by more than 30% is considered a cytotoxic effect [53]. The current study demonstrated that cell viability was reduced after treatment with SDF and SDN by approximately 8% and 13%, respectively.

The limitation of the current study was that the cells were treated with a single concentration of SDF or SDN, which may not represent the clinical situation. The concentration of the solution may be reduced or diluted upon penetration through dentin. It was reported that the concentration of silver ions in dentin was reduced from 1.7 to 0.3 wt % at 20 μm depth [12]. The actual concentration of ions at the pulpal region may be much lower than that used in the current study. Hence, a more relevant model, such as a dentin penetration test, which contains a dentin barrier over the pulp cells, should be used in future studies [54]. Additionally, future work could examine the remineralizing effects of SDN or $AgNO_3$ combined with NaF [17,18]. This could help provide alternative options for materials to control caries when SDF is not available.

5. Conclusions

In this study, we compared the in vitro remineralizing action between silver diamine nitrate (SDN) and silver diamine fluoride (SDF). The use of SDF provided a superior increase in apatite formation compared with SDN. However, the precipitation of silver salts occluding dentinal tubules in demineralized dentin observed with SDF and SDN was comparable. Additionally, the cytotoxic effects on dental pulp cells with SDN were not significantly different compared to SDF. SDN may be considered as an alternative material to control caries. However, more clinical studies are needed to confirm the anti-cavity action of the material.

Supplementary Materials: The following supporting information can be downloaded at: https://www.mdpi.com/article/10.3390/jfb13010016/s1, Video S1: Representative specimen of control group shows no mineral precipitation. Videos S2 and S3: Representative specimens of SDN and SDF groups, respectively. Mineral precipitation is indicated in green.

Author Contributions: Conceptualization, M.K., A.S., S.P., and P.P. (Piyaphong Panpisut); methodology, S.S., P.P. (Phakkhananan Pakawanit), C.R., S.P., and P.P. (Piyaphong Panpisut); software, S.S., P.P. (Phakkhananan Pakawanit), C.R., S.P., and P.P. (Piyaphong Panpisut); validation, M.K., A.S., P.P. (Phakkhananan Pakawanit), C.R., S.P., and P.P. (Piyaphong Panpisut); formal analysis, S.S., P.P. (Phakkhananan Pakawanit), C.R., S.P., and P.P. (Piyaphong Panpisut); investigation, A.S., S.S., P.P. (Phakkhananan Pakawanit), C.R., S.P., and P.P. (Piyaphong Panpisut); resources, M.K., A.S., P.P. (Phakkhananan Pakawanit), C.R., S.P., and P.P. (Piyaphong Panpisut); data curation, S.S., P.P. (Phakkhananan Pakawanit), C.R., S.P., and P.P. (Piyaphong Panpisut); writing—original draft preparation, S.S., P.P. (Phakkhananan Pakawanit), C.R., S.P., and P.P. (Piyaphong Panpisut); writing—review and editing, S.S., P.P. (Phakkhananan Pakawanit), C.R., S.P., and P.P. (Piyaphong Panpisut); visualization, S.S., P.P. (Phakkhananan Pakawanit), C.R., S.P., and P.P. (Piyaphong Panpisut); supervision, M.K., P.P. (Phakkhananan Pakawanit), C.R., and P.P. (Piyaphong Panpisut); project administration, P.P. (Phakkhananan Pakawanit), C.R., S.P., and P.P. (Piyaphong Panpisut); funding acquisition, M.K., A.S., P.P. (Phakkhananan Pakawanit), C.R., S.P., and P.P. (Piyaphong Panpisut). All authors have read and agreed to the published version of the manuscript.

Funding: This research was supported by the Faculty of Dentistry, Thammasat University.

Institutional Review Board Statement: The study was conducted according to the guidelines of the Declaration of Helsinki and approved by the Ethics Review Sub-Committee for Research Involving Human Research Subjects of Thammasat University (protocol code 150/2562, date of approval 2 October 2019).

Informed Consent Statement: Patient consent was waived, as patient identification for the extracted teeth was not recorded.

Data Availability Statement: The datasets generated and/or analyzed during the current study are available from the corresponding author upon reasonable request.

Acknowledgments: The authors are grateful for the support from the Research Unit in Dental and Bone Substitute Biomaterials and the Faculty of Dentistry at Thammasat University, Thammasat University Hospital, and Thammasat University Center of Scientific Equipment for Advanced Research (TUCSEAR). We sincerely appreciate the technical support from the XTM beamline (BL1.2 W) at the Synchrotron Light Research Institute (public organization), Nakhon Ratchasima, Thailand.

Conflicts of Interest: A.S. provided the SDN for the current study. All other authors declare no conflict of interest. The funders had no role in the design of the study; in the collection, analyses, or interpretation of data; in the writing of the manuscript; or in the decision to publish the results.

References

1. Marchini, L.; Ettinger, R.; Hartshorn, J. Personalized Dental Caries Management for Frail Older Adults and Persons with Special Needs. *Dent. Clin.* **2019**, *63*, 631–651. [CrossRef]
2. Bashir, N.Z. Update on the prevalence of untreated caries in the US adult population, 2017–2020. *J. Am. Dent. Assoc.* **2021**, in press. [CrossRef] [PubMed]
3. Schwendicke, F.; Splieth, C.; Breschi, L.; Banerjee, A.; Fontana, M.; Paris, S.; Burrow, M.F.; Crombie, F.; Page, L.F.; Gaton-Hernandez, P.; et al. When to intervene in the caries process? An expert Delphi consensus statement. *Clin. Oral Investig.* **2019**, *23*, 3691–3703. [CrossRef] [PubMed]
4. Wajahat, M.; Abbas, B.; Tariq, K.; Imran, E.; Aslam, S.; Khurshid, Z. Parental perception of silver diamine fluoride for the management of dental caries. *J. Taibah Univ. Med. Sci.* **2022**, in press. [CrossRef]
5. van Strijp, G.; van Loveren, C. No Removal and Inactivation of Carious Tissue: Non-Restorative Cavity Control. *Monogr. Oral Sci.* **2018**, *27*, 124–136. [CrossRef] [PubMed]
6. Johhnson, B.; Serban, N.; Griffin, P.M.; Tomar, S.L. Projecting the economic impact of silver diamine fluoride on caries treatment expenditures and outcomes in young U.S. children. *J. Public Health Dent.* **2019**, *79*, 215–221. [CrossRef]
7. Haq, J.; Khurshid, Z.; Santamaría, R.M.; Abudrya, M.; Schmoeckel, J.; Zafar, M.S.; Splieth, C.H. Silver diamine fluoride: A magic bullet for caries management. *Fluoride* **2021**, *54*, 210–218.
8. Crystal, Y.O.; Niederman, R. Evidence-based dentistry update on silver diamine fluoride. *Dent. Clin.* **2019**, *63*, 45–68. [CrossRef]
9. Zhao, I.S.; Gao, S.S.; Hiraishi, N.; Burrow, M.F.; Duangthip, D.; Mei, M.L.; Lo, E.C.; Chu, C.H. Mechanisms of silver diamine fluoride on arresting caries: A literature review. *Int. Dent. J.* **2018**, *68*, 67–76. [CrossRef]
10. Mei, M.L.; Lo, E.C.M.; Chu, C.H. Arresting dentine caries with silver diamine fluoride: What's behind it? *J. Dent. Res.* **2018**, *97*, 751–758. [CrossRef]
11. Sayed, M.; Hiraishi, N.; Matin, K.; Abdou, A.; Burrow, M.F.; Tagami, J. Effect of silver-containing agents on the ultra-structural morphology of dentinal collagen. *Dent. Mater.* **2020**, *36*, 936–944. [CrossRef]
12. Willershausen, I.; Schulte, D.; Azaripour, A.; Weyer, V.; Briseno, B.; Willershausen, B. Penetration Potential of a Silver Diamine Fluoride Solution on Dentin Surfaces. An Ex Vivo Study. *Clin. Lab.* **2015**, *61*, 1695–1701. [CrossRef] [PubMed]
13. Kim, S.; Nassar, M.; Tamura, Y.; Hiraishi, N.; Jamleh, A.; Nikaido, T.; Tagami, J. The effect of reduced glutathione on the toxicity of silver diamine fluoride in rat pulpal cells. *J. Appl. Oral Sci.* **2021**, *29*, e20200859. [CrossRef] [PubMed]
14. Fancher, M.E.; Fournier, S.; Townsend, J.; Lallier, T.E. Cytotoxic effects of silver diamine fluoride. *Am. J. Dent.* **2019**, *32*, 152–156. [PubMed]
15. Zhi, Q.H.; Lo, E.C.; Kwok, A.C. An in vitro study of silver and fluoride ions on remineralization of demineralized enamel and dentine. *Aust. Dent. J.* **2013**, *58*, 50–56. [CrossRef]
16. Gao, S.S.; Duangthip, D.; Wong, M.C.M.; Lo, E.C.M.; Chu, C.H. Randomized Trial of Silver Nitrate with Sodium Fluoride for Caries Arrest. *JDR Clin. Trans. Res.* **2019**, *4*, 126–134. [CrossRef]
17. Gao, S.S.; Chen, K.J.; Duangthip, D.; Wong, M.C.M.; Lo, E.C.M.; Chu, C.H. Arresting early childhood caries using silver and fluoride products-A randomised trial. *J. Dent.* **2020**, *103*, 103522. [CrossRef]
18. Zhao, I.S.; Mei, M.L.; Li, Q.L.; Lo, E.C.M.; Chu, C.H. Arresting simulated dentine caries with adjunctive application of silver nitrate solution and sodium fluoride varnish: An in vitro study. *Int. Dent. J.* **2017**, *67*, 206–214. [CrossRef]
19. Gonzalez-Cabezas, C.; Fernandez, C.E. Recent Advances in Remineralization Therapies for Caries Lesions. *Adv. Dent. Res.* **2018**, *29*, 55–59. [CrossRef]

20. Gao, S.S.; Zhao, I.S.; Duffin, S.; Duangthip, D.; Lo, E.C.M.; Chu, C.H. Revitalising silver nitrate for caries management. *Int. J. Environ. Res. Public Health* **2018**, *15*, 80. [CrossRef]
21. Saito, T.; Toyooka, H.; Ito, S.; Crenshaw, M.A. In vitro study of remineralization of dentin: Effects of ions on mineral induction by decalcified dentin matrix. *Caries Res.* **2003**, *37*, 445–449. [CrossRef] [PubMed]
22. Hu, S.; Meyer, B.; Duggal, M. A silver renaissance in dentistry. *Eur. Arch. Paediatr. Dent.* **2018**, *19*, 221–227. [CrossRef] [PubMed]
23. Jennings, M.C.; Minbiole, K.P.; Wuest, W.M. Quaternary Ammonium Compounds: An Antimicrobial Mainstay and Platform for Innovation to Address Bacterial Resistance. *ACS Infect. Dis.* **2015**, *1*, 288–303. [CrossRef] [PubMed]
24. Chen, K.J.; Gao, S.S.; Duangthip, D.; Lo, E.C.M.; Chu, C.H. The caries-arresting effect of incorporating functionalized tricalcium phosphate into fluoride varnish applied following application of silver nitrate solution in preschool children: Study protocol for a randomized, double-blind clinical trial. *Trials* **2018**, *19*, 352. [CrossRef] [PubMed]
25. Fraihat, N.; Madae'en, S.; Bencze, Z.; Herczeg, A.; Varga, O. Clinical Effectiveness and Cost-Effectiveness of Oral-Health Promotion in Dental Caries Prevention among Children: Systematic Review and Meta-Analysis. *Int. J. Environ. Res. Public Health* **2019**, *16*, 2668. [CrossRef] [PubMed]
26. Srisomboon, S.; Kettratad, M.; Pakawanit, P.; Rojviriya, C.; Phantumvanit, P.; Panpisut, P. Effects of Different Application Times of Silver Diamine Fluoride on Mineral Precipitation in Demineralized Dentin. *Dent. J.* **2021**, *9*, 70. [CrossRef]
27. Chen, Z.; Cao, S.; Wang, H.; Li, Y.; Kishen, A.; Deng, X.; Yang, X.; Wang, Y.; Cong, C.; Wang, H.; et al. Biomimetic remineralization of demineralized dentine using scaffold of CMC/ACP nanocomplexes in an in vitro tooth model of deep caries. *PLoS ONE* **2015**, *10*, e0116553. [CrossRef]
28. Sayed, M.; Matsui, N.; Hiraishi, N.; Inoue, G.; Nikaido, T.; Burrow, M.F.; Tagami, J. Evaluation of discoloration of sound/demineralized root dentin with silver diamine fluoride: In-vitro study. *Dent. Mater. J.* **2019**, *38*, 143–149. [CrossRef]
29. *BS ISO 23317:2014*; Implants for surgery. In vitro evaluation for apatite-forming ability of implant materials. BSI Standards Limited: London, UK, 2014.
30. Kokubo, T.; Yamaguchi, S. Simulated body fluid and the novel bioactive materials derived from it. *J. Biomed. Mater Res. A* **2019**, *107*, 968–977. [CrossRef] [PubMed]
31. Dridi, A.; Riahi, K.Z.; Somrani, S. Mechanism of apatite formation on a poorly crystallized calcium phosphate in a simulated body fluid (SBF) at 37 °C. *J. Phys. Chem. Solids* **2021**, *156*, 110122. [CrossRef]
32. Liu, Y.; Yao, X.; Liu, Y.W.; Wang, Y. A Fourier transform infrared spectroscopy analysis of carious dentin from transparent zone to normal zone. *Caries Res.* **2014**, *48*, 320–329. [CrossRef] [PubMed]
33. Zhang, Y.; Wang, Z.; Jiang, T.; Wang, Y. Biomimetic regulation of dentine remineralization by amino acid in vitro. *Dent. Mater.* **2019**, *35*, 298–309. [CrossRef] [PubMed]
34. Lopes, C.d.C.A.; Limirio, P.H.J.O.; Novais, V.R.; Dechichi, P. Fourier transform infrared spectroscopy (FTIR) application chemical characterization of enamel, dentin and bone. *Appl. Spectrosc. Rev.* **2018**, *53*, 747–769. [CrossRef]
35. Yin, I.X.; Yu, O.Y.; Zhao, I.S.; Mei, M.L.; Li, Q.L.; Tang, J.; Lo, E.C.M.; Chu, C.H. Inhibition of dentine caries using fluoride solution with silver nanoparticles: An in vitro study. *J. Dent.* **2020**, *103*, 103512. [CrossRef]
36. Berzina-Cimdina, L.; Borodajenko, N. Research of calcium phosphates using Fourier transform infrared spectroscopy. In *Infrared Spectroscopy-Materials Science, Engineering and Technology*; InTech: Rijeka, Croatia, 2012; pp. 123–148.
37. Vidal Bde, C.; Mello, M.L. Collagen type I amide I band infrared spectroscopy. *Micron* **2011**, *42*, 283–289. [CrossRef]
38. Vlassenbroeck, J.; Masschaele, B.; Cnudde, V.; Dierick, M.; Pieters, K.; Van Hoorebeke, L.; Jacobs, P. Octopus 8: A High performance tomographic reconstruction package for X-ray tube and synchrotron micro-CT. In *Advances in X-ray Tomography for Geomaterials*; John Wiley & Sons: Hoboken, NJ, USA, 2006; pp. 167–173.
39. Limaye, A. Drishti: A volume exploration and presentation tool. In Proceedings of the SPIE8506, Developments in X-Ray Tomography VIII, San Diego, CA, USA, 17 October 2012; Volume 8506, pp. 85061–85069. [CrossRef]
40. Mungpayabarn, H.; Patntirapong, S. Timing of geranylgeraniol addition increases osteoblast activities under alendronate condition. *J. Oral Biol. Craniofac. Res.* **2021**, *11*, 396–401. [CrossRef]
41. Lilakhunakon, C.; Suwanpateeb, J.; Patntirapong, S. Inhibitory Effects of Alendronate on Adhesion and Viability of Preosteoblast Cells on Titanium Discs. *Eur. J. Dent.* **2021**, *15*, 502–508. [CrossRef]
42. Panpisut, P.; Suppapatpong, T.; Rattanapan, A.; Wongwarawut, P. Monomer conversion, biaxial flexural strength, apatite forming ability of experimental dual-cured and self-adhesive dental composites containing calcium phosphate and nisin. *Dent. Mater. J* **2021**, *40*, 399–406. [CrossRef]
43. Clift, F. Artificial methods for the remineralization of hydroxyapatite in enamel. *Mater. Today Chem.* **2021**, *21*, 100498. [CrossRef]
44. Iijima, M.; Onuma, K. Roles of Fluoride on Octacalcium Phosphate and Apatite Formation on Amorphous Calcium Phosphate Substrate. *Cryst. Growth Des.* **2018**, *18*, 2279–2288. [CrossRef]
45. Lin, M.; Luo, Z.Y.; Bai, B.F.; Xu, F.; Lu, T.J. Fluid mechanics in dentinal microtubules provides mechanistic insights into the difference between hot and cold dental pain. *PLoS ONE* **2011**, *6*, e18068. [CrossRef] [PubMed]
46. Vogel, G.L. Oral fluoride reservoirs and the prevention of dental caries. *Monogr. Oral Sci.* **2011**, *22*, 146–157. [CrossRef] [PubMed]
47. Mei, M.L.; Nudelman, F.; Marzec, B.; Walker, J.M.; Lo, E.C.M.; Walls, A.W.; Chu, C.H. Formation of fluorohydroxyapatite with silver diamine fluoride. *J. Dent. Res.* **2017**, *96*, 1122–1128. [CrossRef] [PubMed]
48. Lou, Y.L.; Botelho, M.G.; Darvell, B.W. Reaction of silver diamine [corrected] fluoride with hydroxyapatite and protein. *J. Dent.* **2011**, *39*, 612–618. [CrossRef] [PubMed]

49. Mei, M.L.; Ito, L.; Cao, Y.; Li, Q.L.; Lo, E.C.; Chu, C.H. Inhibitory effect of silver diamine fluoride on dentine demineralisation and collagen degradation. *J. Dent.* **2013**, *41*, 809–817. [CrossRef]
50. Zhao, I.S.; Mei, M.L.; Burrow, M.F.; Lo, E.C.; Chu, C.H. Effect of silver diamine fluoride and potassium iodide treatment on secondary caries prevention and tooth discolouration in cervical glass ionomer cement restoration. *Int. J. Mol. Sci.* **2017**, *18*, 340. [CrossRef]
51. Romero, M.; Lippert, F. Indirect caries-preventive effect of silver diamine fluoride on adjacent dental substrate: A single-section demineralization study. *Eur. J. Oral. Sci.* **2021**, *129*, e12751. [CrossRef]
52. Reyes-Gasga, J.; Martinez-Pineiro, E.L.; Rodriguez-Alvarez, G.; Tiznado-Orozco, G.E.; Garcia-Garcia, R.; Bres, E.F. XRD and FTIR crystallinity indices in sound human tooth enamel and synthetic hydroxyapatite. *Mater. Sci. Eng. C Mater. Biol. Appl.* **2013**, *33*, 4568–4574. [CrossRef]
53. *10993–5: 2009*; Biological evaluation of medical devices. Part 5: Tests for in vitro cytotoxicity. BSI Standards Limited: London, UK, 2009.
54. Hadjichristou, C.; About, I.; Koidis, P.; Bakopoulou, A. Advanced in Vitro Experimental Models for Tissue Engineering-based Reconstruction of a 3D Dentin/pulp Complex: A Literature Review. *Stem Cell Rev. Rep.* **2021**, *17*, 785–802. [CrossRef]

Journal of
Functional
Biomaterials

MDPI

Article

Dental Pulp Response to Different Types of Calcium-Based Materials Applied in Deep Carious Lesion Treatment—A Clinical Study

Antoanela Covaci [1,2], Lucian Toma Ciocan [3,*], Bogdan Gălbinașu [3,*], Mirela Veronica Bucur [3], Mădălina Matei [2] and Andreea Cristiana Didilescu [1]

1 Department of Embryology, Faculty of Dental Medicine, "Carol Davila" University of Medicine and Pharmacy, 050474 Bucharest, Romania; antoanela.covaci@ugal.ro (A.C.); Andreea.Didilescu@umfcd.ro (A.C.D.)
2 Department of Dental Medicine, Faculty of Medicine and Pharmacy, Dunarea de Jos University of Galati, 800010 Galati, Romania; Madalina.Matei@ugal.ro
3 Department of Prosthetics Technology and Dental Materials, Faculty of Dental Medicine, "Carol Davila" University of Medicine and Pharmacy, 010221 Bucharest, Romania; mirela.bucur@umfcd.ro
* Correspondence: lucian.ciocan@umfcd.ro (L.T.C.); bogdan.galbinasu@umfcd.ro (B.G.)

Abstract: Dental pulp vitality preservation in dental caries treatment is a major goal in odontotherapy. The main objective of this study was to compare dental pulp tissue responses to vital therapies in deep carious lesions, using different calcium-based materials. An ambispective study was conducted on 47 patients. Ninety-five teeth with deep carious lesions were treated. Among them, 25 (26.32%) were diagnosed with pulpal exposures and treated by direct pulp capping. Indirect pulp capping was applied when pulp exposure was absent (n = 70; 73.68%). Fifty teeth (52.63%) were treated with TheraCal LC (prospective study), 31 teeth (32.63%) with Calcimol LC, and 14 teeth (14.74%) with Life Kerr AC (retrospective study). The results show that the survival rate for dental pulp was 100% for Life Kerr AC, 92% for TheraCal LC, and 83.87% for Calcimol LC, without significant differences. Apparently, self-setting calcium hydroxide material provided better dental pulp response than the two light-cured materials, regardless of their composition, that is, either calcium -hydroxide or calcium silicate-based. We will need a significant number of long-term clinical studies with the highest levels of evidence to determine the most adequate biomaterials for vital pulp therapies.

Keywords: pulp capping; dental; calcium hydroxide; silicates

1. Introduction

Dental pulp vitality preservation in dental caries treatment is a major goal in odontotherapy. Both modern dentistry and extended histological research conclude and accept that nothing can replace the dental pulp, generating the same benefits for the tooth and the periodontium [1].

Dental caries is a disease caused by multiple factors and involves interactions of three factors: the body of the host, represented primarily by the teeth and the saliva; the diet, determined by the availability of fermentable carbohydrates; and the microbiota, which are acid-producing bacteria [2].

Depending on their position, depth, or extension, caries generates major difficulties in achieving the correct technique of classic treatment. Sometimes, the occult evolution (by the interproximal position) in the absence of a careful clinical/radiological exam, makes pathology detection impossible before the irreversible pulp disease. Rapidly progressive or slowly progressive lesions inevitably converge toward pulp damage, so, in both situations, the treatment must be rapid, complex, and rigorous [3].

Ever since the mid-1970s, studies have indicated that the pulp tissue can tolerate different dental restorative materials as long as bacteria and their toxins can be excluded from the pulp tissue [4]. This is the goal of direct/indirect capping.

Calcium hydroxide was introduced in dentistry in 1921 and has been considered the "gold standard" of direct pulp-capping materials for many years. [5] This material is considered to have excellent antibacterial properties [6], and one of the studies found a complete reduction of the micro-organisms that are frequently associated with pulp infections after only one hour of contact with calcium hydroxide [7]. More than that, calcium hydroxide has one of the best clinical success rates and long-term follow-up rates as a pulp-capping agent after different periods, even after 10 years [8].

But calcium hydroxide is not infallible. The self-cure formulations are highly soluble and can dissolve in time [9], but it has been noticed that by the time the calcium hydroxide disappears because it dissolves, new bridges of detin are formed [8,10]. It provides a poor seal [11] and has no inherent adhesive qualities. Another concern about this material would be the appearance of "tunnel defects" in reparative dentin formed underneath calcium hydroxide pulp caps [12].

MTA is primarily composed of calcium oxide in dicalcium silicate, tricalcium silicate, and tricalcium aluminate form. Bismuth oxide can be added for its radiopacity effect [13]. MTA is considered a silicate cement rather than an oxide mixture, and so its biocompatibility is based on its reaction products [14]. It is important to note that the primary reaction product of MTA with water is calcium hydroxide [15], and so calcium hydroxide's formation is actually the one that provides MTA's biocompatibility, so they are rather similar. However, a significant difference would be the fact that MTA provides some seal to tooth structure [16].

An important downside to MTA manipulation and clinical use is the prolonged setting time; some products need more than 2 h [17]. This implies that pulp capping with MTA is clinically possible either using a quick-setting liner to protect the MTA during permanent restoration placement or performing a two-step procedure.

The main objective of the present study was to compare the clinical and biological effects of different calcium-based pulp-capping materials on dental pulp responses to vital therapies in deep carious lesions. The specific aim was to clinically assess and compare the pulp vitality, following the above-mentioned therapies.

2. Materials and Methods

This ambispective clinical study was conducted in compliance with the research ethics legislation currently in place in Romania. Informed consent was obtained from all subjects involved in the study. The study was conducted in accordance with the Declaration of Helsinki, and the protocol was approved by the Ethics Committee of Dunarea de Jos University of Galati (no. 4842/20/02/2020). The study was conducted by the same investigator (Antoanela Covaci).

2.1. Sample Selection

The medical files of patients who underwent direct or indirect pulp-capping therapies in Dr. Antoanela Covaci's private practice from February 2017 to May 2019 consisted of the retrospective, control sample. Patients attending the private practice from 1 July 2020 to 20 December 2021 were included in the prospective study. Patients with direct/indirect pulp cappings with three dentinogenesis-inducing materials, Theracal LC (Bisco Inc., Schaumburg, IL, USA), Calcimol LC (Voco GmbH, Cuxhaven, Germany), and Life Kerr AC (Kerr, Orange, CA, USA), were selected to be included into the study.

Inclusion Criteria

The included patients met certain criteria, namely, thermal sensitivity response compatible with a diagnosis of tooth vitality and association of radiographic examination, a reasonable state of health and oral hygiene, and no associated periodontal pathology. The included treatments had the same protocol: isolation of the operative field with cotton rolls type 2 and cavity cleaned with neophaline. The hydrogen peroxide on sterilized cotton pellets was used for hemostasis in case of accidental pulp exposure. Teeth in which restorations were performed using self-etching adhesive systems such as Filltec, GC, Voco, Beautifil, and Flow were included. Light-cured composite resins were used as restorative materials in all the therapies.

2.2. Evaluation of the Clinical Procedures of Direct/Indirect Pulp Capping

Ninety-five direct/indirect pulp-capping therapies performed on 47 patients were included and divided into three groups: the first group with 50 teeth was treated with TheraCal LC (13 anterior teeth, 16 premolars, and 21 molars); the second group with 31 teeth was treated with Calcimol LC (12 anterior teeth, 9 premolars, and 10 molars), and the third group with 14 teeth was treated with Life Kerr AC (1 anterior tooth and 13 molars). The etiologies of the pulp exposures were different because of extensive carious lesions and accidentally, at the moment of removal, of soft, infiltrated dentin; therefore, different dimensions of the exposure sites were reported. In all clinical cases, the therapies were performed with one of the above-mentioned biomaterials, followed by the placement of direct resin or glass-ionomer restorations of the crowns. The permanent restorations were performed at the appointment in which the direct/indirect pulp capping was performed.

2.3. Evaluation of the Teeth after Direct/Indirect Capping Therapy

One month, 3 months, and 6 months after treatment, clinical controls were performed. During these periodical controls, qualitative tests were performed such as the pulp vitality test, which included the vertical and horizontal percussion tests (see Figure 1).

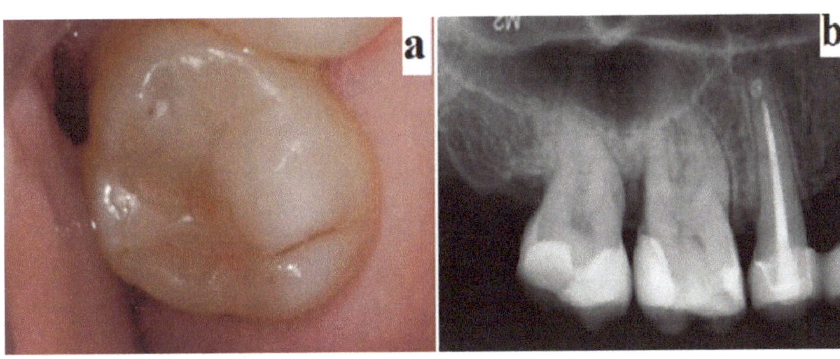

Figure 1. Example of evaluation 6 months post-op: Clinical evaluation—marginal infiltration (**a**); periapical X-ray evaluation (**b**).

Therapies on teeth that remained asymptomatic, with normal sensitivity tests and no other radiographic signs, such as periapical pathology, were considered as clinical successes.

2.4. Statistical Analysis

Data distributions were expressed as means, standard deviations (SD) and percentages. Pearson's chi-squared tests were used for categorical measures. When the expected frequency of any cell in the table was <5, Fisher's exact test was used

Statistical analyses were performed using Stata/IC 16 (StataCorp. 2019. Stata Statistical Software: Release 16. StataCorp LLC.: College Station, TX, USA), and p-values < 0.05 were considered statistically significant.

3. Results

The study included 47 urban patients (70.21% females; mean age 34.66 ± 11.15 years). Ninety-five teeth, presenting deep carious lesions, were treated. Among them, 25 (26.32%)

were diagnosed with pulpal exposures and treated by direct pulp capping. Indirect pulp capping was performed when pulp exposure was absent. Fifty teeth (52.63%) were treated with TheraCal LC (prospective study); 31 teeth (32.63%) were treated with Calcimol LC, and 14 teeth (14.74%) were treated with Life Kerr AC (retrospective study). Regarding direct pulp capping, 4 teeth were treated with Life Kerr AC, and 21 teeth were treated with Theracal LC. Indirect pulp capping was applied as follows: 10 teeth with Life Kerr AC, 31 teeth with Calcimol, and 29 teeth with TheraCal LC.

Features of the 95 carious lesions treated and analyzed are presented in Table 1. The prospective cohort comprised significantly more pulp exposures than the retrospective cohort (Table 1).

Table 1. Characteristics of the carious lesions treated.

	Retrospective Cohort (n = 45)		Prospective Cohort (n = 50)		p
	n	%	n	%	
Pulpal exposure					
Yes	4	8.89	21	42	<0.001
No	41	91.11	29	58	
Capping material					
TheraCal LC	0	0	50	100	<0.001
Calcimol LC	31	68.89	0	0	
Life Kerr AC	14	31.11	0	0	
Affected teeth					
Anterior	13	28.89	13	26	0.753
Posterior	32	71.11	37	74	
Lesion activity					
Active	4	8.89	0	0	0.572
Arrested	41	91.11	50	100	

The results of tooth vitality preservation 6 months after treatment for the materials taken into the study are presented in Table 2.

Table 2. Outcome assessment.

Capping Material	Vitality Preservation		p
	No	Yes	
TheraCal LC	4 (8%)	46 (92%)	0.236
Calcimol LC	5 (16.13%)	26 (83.87%)	
Life Kerr AC	0	14 (100%)	

4. Discussion

Taking into account the analysis of the results registered in the present study, we can draw multiple conclusions.

Regarding the specific aim of the study, the rate of preservation of pulp vitality was different for the three pulp-capping materials although the difference was not statistically significant (see Table 2). Among the only 14 teeth (15% of 95 teeth analyzed) treated with Life Kerr AC, none of them lost vitality after 6 months. For the other two materials, the vitality of the treated teeth was lost in 8% of the cases treated with TheraCal LC and in 16.13% of the cases treated with Calcimol LC.

The above-mentioned results were not related to the position of the affected tooth (anterior or posterior), and they seem not even related to the initial carious activity of the treated lesion. All materials for pulp capping taken into the study (TheraCal LC, Calcimol LC, and Life Kerr AC) are indicated to be used for pulp capping, although Calcimol LC is recommended only for indirect pulp-capping treatments.

At the time of cavity preparation, by getting close to pulp tissue or, even worse, exposing it using rotary instruments, the pulp tissues can become inflamed or necrotized. In this case, the clinician often makes the decision to perform an endodontic treatment. For this reason, the materials indicated for pulp capping should act as a barrier and protect the vitality of the entire pulp by covering the minimally exposed tissue and preventing the need for further endodontic treatments. At the same time, the capping material used should prompt a regenerative response from the host side [18].

TheraCal LC (Bisco Inc., Schaumburg, IL, USA) is a light-cured, resin–calcium silicate matrix. This liner material is recommended for direct and indirect pulp capping. It contains monomers of polymerizable methacrylate, Portland cement type III, polyethylene glycol dimethacrylate, and barium zirconate [19].

Calcimol LC (Voco GmbH, Cuxhaven, Germany) is a light-cured, pulp-capping material. This material represents a resin-modified calcium ion-releasing liner. In this material, the calcium dihydroxide is embedded in a resin–polymethacrylate matrix: urethane dimethacrylate, dimethylaminoethyl-methacrylate, and triethyleneglycol dimethacrylate (TEGDMA) [20].

Life Kerr AC (Kerr GmbH, Karlsruhe, Germany) is a self-setting, calcium ion-releasing liner and pulp-capping material. It contains calcium dihydroxide, N-ethyl-o(or p)-toluenesulphonamide, zinc and calcium oxide, methyl salicylate, and 2,2-dimethylpropane-1,3-diol [21].

Light-curable, resin-modified calcium hydroxide materials, such as two of the materials taken into the study, are TheraCal LC (Bisco) and Calcimol LC (Voco), largely used for direct pulp capping. Compared to the conventional two-paste calcium hydroxide systems, the third material taken into the study, Life Kerr (Kerr), the resin-modified versions have several advantages, including ease of handling, light polymerization, and superior physical properties. They are also minimally affected by phosphoric acid and have low water solubility, which means that they do not dissolve in time.

Despite the multiple advantages of the resin-modified calcium hydroxide liners, due to the light-activated polymerization, we can say that there is a significant risk of free residual monomers left at the pulp-capping site. It is known that unpolymerized monomers are toxic to pulp cells [22]. For example, Calcimol LC was reported to present higher cytotoxicity to MDPC-23 cells than another resin-free calcium hydroxide paste. However, the composite resin is considered to present mild to no toxic effects to the odontoblast-like MDPC-23 cells if it is polymerized [23]. Another study shows that resin-modified calcium hydroxide is not more cytotoxic than the control calcium hydroxide. When light resin-modified pulp-capping materials with light-activated polymerization are sufficiently cured, with a longer curing time, the cytotoxicity effect of the resin disappears. Meanwhile, in the time of polymerization, $OH-$ is released and can cause some cytotoxicity [24]. We noticed that a mean rate of 70% conversion of the polymerization in the case of dimethacrylate monomers does not mean the presence of 30% unreacted free monomers. It means that only 30% of the methacrylate groups remains still active for polymerization, but among those, most of them are already inside the polymer matrix. Overall, only a small percentage (9% of monomers) can be considered free. (i.e., both methacrylate groups in one monomer are not cured), and most of these free monomers are located inside the polymer matrix (cannot be released) [25]. There are some studies [26,27] that report that cytotoxicity was not observed in the MG63 cells treated with TheraCal LC, and after 5 days, the cells are organized as a confluent monolayer as demonstrated by fluorescence microscopy observations. Theracal LC shows biocompatibility on MG63 cells allowing physiological cell growth and differentiation. Chemical and physical properties and Theracal LC biocompatibility observed in in vitro studies still consider this cement as an efficient pulp-capping material for the vital pulp therapy [27].

From a clinical point of view, there is a difference between the two resin-modified pulp-capping materials taken into the study. Calcimol LC is easier to handle and more tooth-colored compared to the opaque white of TheraCal LC. In addition, taking into consideration the manufacturer's indications, Calcimol LC can be used with or without dental adhesives [28,29]. Materials with new compositions are evaluated comprehensively before their clinical application. There are recent studies that examined whether the lower calcium ion-releasing ability, together with the cytotoxicity because of unpolymerized resin monomers of resin-modified calcium ion-releasing liners, has an influence on its biological and clinical performance [30]. The pH values evaluated in another study [31] were slightly alkaline for TheraCal LC and Calcimol LC, compared to the control group. The lower pH values registered were because the ions were dispersed through the different dentine thicknesses that remained. This discovery contrasts with other studies [32,33]. Approaching a physiological pH within 60 days during this study may provide a positive environment for pulpal cell viability and metabolic movement with the reparative dentine development. TheraCal LC has been demonstrated to discharge higher Ca2+ ions right after application and to make a natural pH, very close to the physiological one in the first two months. Further clinical tests are required to measure the release of different biologically active ions from TheraCal LC, which can contribute to the clinical success of these materials in vital pulp therapies [31,32,34].

Another randomized systematic review study, analyzing long-term clinical and radiographic evaluation of the effectiveness of direct pulp capping materials, showed that multiple variables must be taken into consideration for an accurate interpretation of pulp-capping material's efficiency [35]. The aim of this review was to assess the effectiveness of 12 different direct pulp capping materials for dental pulp exposures. Long-term clinical and paraclinical (X-rays) evaluations of the efficiency of different direct pulp-capping materials used on teeth with pulp exposure were included. After a risk of bias assessment and data acquisition and interpretation from 496 identified articles, only 15 met the eligibility criteria. From all the studies that were included in those articles, a total of 1322 teeth were treated with 12 different types of direct pulp-capping materials. However, the results were based on the present studies, which were all judged to have a high risk of misinterpretation. In this evaluation, many materials were studied, and some of them seemed to perform better than calcium hydroxide materials, as for example Life Kerr (Kerr). However, unlike calcium hydroxide, all the other materials were supported by only a small number of studies. Therefore, more long-term clinical and radiographic studies with lower a risk of bias are needed [35].

Although calcium hydroxide (as Life Kerr) has long been considered the gold standard for direct pulp capping, it has some disadvantages: The high pH can irritate the dental pulp and can cause the inflammation or the necrosis of the exposed pulp surface. The newly formatted dentine can have tunnel-shaped defects, and the dissolution in time may lead to failure of the long-term seal. All these disadvantages are probably responsible for the wide differences in success rates, ranging from 52% to 100% [35–37].

The differences in the protocols of isolation and antiseptisation are an important factor that must be taken into consideration for the accurate interpretation of different studies. In another study with similar conditions of isolation and antiseptisation, a study in which 69 teeth were treated with calcium hydroxide, 57 received indirect pulp capping, with 53 (93.0%) showing a successful outcome and 4 (7.0%) an unsuccessful outcome [38]. This result can explain the efficiency of 100% of calcium hydroxide that was analyzed in our study and contributed to the smaller number of teeth (only 14) involved.

The limits of the study include the design and relatively small sample size. Because of the restrictions imposed by COVID-19, patient recruitment availability was seriously affected, and we could not provide similar sample sizes for the materials that were investigated. It is obvious that the retrospective component of the study relied mostly on the accurate recordkeeping. However, considering that the same investigator was involved in all treatments, we consider that the results of the study were less biased.

More than 20 types of biomaterials are effective in direct pulp capping. Until now, an ideal pulpotomy material has not been established [39–41]. A greater number of long-term clinical studies with highest levels of evidence (randomized control tests) are required to determine the best composition biomaterial for direct or indirect pulp capping.

5. Conclusions

Within the limits of the study, our results suggest a better preservation of dental pulp vitality in the case of self-setting calcium hydroxide Life Kerr AC (Kerr) as compared to resin-modified calcium-releasing TheraCal LC (Bisco) and Calcimol LC (Voco). Among the two light-activated materials, TheraCal LC (Bisco) and Calcimol LC (Voco), TheraCal LC seems to have a better potential in keeping pulp vitality, probably due to a better local pH maintenance.

A greater number of long-term clinical studies with the highest levels of evidence are needed to determine the most adequate biomaterials for vital pulp therapies. Moreover, extensive research has to be carried out to improve dental materials in order to maintain the dental pulp potential and facilitate its regeneration in the case of cavities aggression.

Author Contributions: Conceptualization, A.C., L.T.C., B.G. and A.C.D.; methodology, A.C., L.T.C., B.G. and A.C.D.; validation, A.C., M.M. and A.C.D.; formal analysis, B.G., M.V.B. and A.C.D.; investigation, A.C. and M.M.; writing—original draft preparation, A.C., A.C.D., B.G. and M.M.; writing—review and editing, L.T.C., B.G. and A.C.D.; visualization, A.C., L.T.C., B.G., M.V.B., M.M. and A.C.D. All authors have read and agreed to the published version of the manuscript.

Funding: This research received no external funding.

Institutional Review Board Statement: The study was conducted in accordance with the Declaration of Helsinki, and the protocol was approved by the Ethics Committee of Dunarea de Jos University of Galati (no. 4842/ 20/02/2020).

Informed Consent Statement: Informed consent was obtained from all subjects involved in the study.

Data Availability Statement: The data presented in this study are available on request from the corresponding author.

Conflicts of Interest: The authors declare no conflict of interest.

References

1. Zhang, W.; Yelick, P.C. Vital pulp therapy-current progress of dental pulp regeneration and revascularization. *Int. J. Dent.* **2010**, *2010*, 856087. [CrossRef]
2. Rathee, M.; Sapra, A. Dental Caries. In *StatPearls*; StatPearls Publishing: Treasure Island, FL, USA, 2022. Available online: https://www.ncbi.nlm.nih.gov/books/NBK551699/ (accessed on 19 March 2022).
3. Hashem, D.; Mannocci, F.; Patel, S.; Manoharan, A.; Brown, J.E.; Watson, T.F.; Banerjee, A. Clinical and radiographic assessment of the efficacy of calcium silicate indirect pulp capping: A randomized controlled clinical trial. *J. Dent. Res.* **2015**, *94*, 562–568. [CrossRef] [PubMed]
4. Hilton, T.J. Keys to clinical success with pulp Capping: A Review of the literature. *Oper. Dent.* **2009**, *34*, 615–625. [CrossRef] [PubMed]
5. Baume, L.J.; Holz, J. Long term clinical assessment of direct pulp capping. *Int. Dent. J.* **1981**, *31*, 251–260. [PubMed]
6. Barthel, C.R.; Levin, L.G.; Reisner, H.M.; Trope, M. TNF-alpha release in monocytes after exposure to calcium hydroxide treated Escherichia coli LPS. *Int. Endod. J.* **1997**, *30*, 155–159. [CrossRef] [PubMed]
7. Stuart, K.G.; Miller, C.H.; Brown, C.E., Jr.; Newton, C.W. The comparative antimicrobial effect of calcium hydroxide. *Oral Surg. Oral Med. Oral Pathol.* **1991**, *72*, 101–104. [CrossRef]
8. Accorinte, M.; Reis, A.; Loguercio, A.; de Araújo, V.; Muench, A. Influence of rubber dam isolation on human pulp responses after capping with calcium hydroxide and an adhesive system. *Quintessence Int.* **2006**, *37*, 205–212.
9. Prosser, H.J.; Groffman, D.M.; Wilson, A.D. The effect of composition on the erosion properties of calcium hydroxide cements. *J Dent Res.* **1982**, *61*, 1431–1435. [CrossRef]
10. Accorinte, M.; Loguercio, A.; Reis, A.; Carneiro, E.; Grande, R.; Murata, S.; Holland, R. Response of human dental pulp capped with MTA and calcium hydroxide powder. *Oper Dent.* **2008**, *33*, 488–495. [CrossRef]
11. Ferracane, J. *Materials in Dentistry, Principles and Applications*, 2nd ed.; Lippincott, Williams & Wilkins: Philadelphia, PA, USA, 2001; pp. 63–64.
12. Kitasako, Y.; Ikeda, M.; Tagami, J. Pulpal responses to bacterial contamination following dentin bridging beneath hard-setting calcium hydroxide and self-etching adhesive resin system. *Dent. Traumatol.* **2008**, *24*, 201–206. [CrossRef]
13. Camilleri, J. Characterization of hydration products of mineral trioxide aggregate. *Int. Endod. J.* **2008**, *41*, 408–417. [CrossRef] [PubMed]
14. Camilleri, J.; Pitt Ford, T.R. Mineral trioxide aggregate: A review of the constituents and biological properties of the material. *Int. Endod. J.* **2006**, *39*, 747–754. [CrossRef] [PubMed]

15. Fridland, M.; Rosado, R. MTA solubility: A long term study. *J. Endod.* **2005**, *31*, 376–379. [CrossRef] [PubMed]
16. Luketić, S.F.; Malcić, A.; Jukić, S.; Anić, I.; Segović, S.; Kalenić, S. Coronal microleakage of two root-end filling materials using a polymicrobial marker. *J. Endod.* **2008**, *34*, 201–203. [CrossRef] [PubMed]
17. Islam, I.; Chng, H.K.; Yap, A.U. Comparison of the physical and mechanical properties of MTA and portland cement. *J. Endod.* **2006**, *32*, 193–197. [CrossRef]
18. Poggio, C.; Arciola, C.R.; Beltrami, R.; Monaco, A.; Dagna, A.; Lombardini, M.; Visai, L. Cytocompatibility and antibacterial properties of capping materials. *Sci. World J.* **2014**, 181945. [CrossRef]
19. BISCO Dental: Dental Adhesive and Dental Cement Products. Available online: http://www.bisco.com (accessed on 20 March 2022).
20. Safety Data Sheet. Available online: https://www.voco.dental/us/portaldata/1/resources/products/safety-data-sheets/us/calcimol-lc_sds_us.pdf (accessed on 21 March 2022).
21. Kerr Dental. Available online: https://www.kerrdental.com/kerr-restoratives/life-pulp-capping-material#docs (accessed on 20 March 2022).
22. Jontell, M.; Hanks, C.T.; Bratell, J.; Bergenholtz, G. Effects of unpolymerized resin components on the function of accessory cells derived from the rat incisor pulp. *J. Dent. Res.* **1995**, *74*, 1162–1167. [CrossRef]
23. Aranha, A.M.; Giro, E.M.; Hebling, J.; Lessa, F.C.; Costa, C.A. Effects of light-curing time on the cytotoxicity of a restorative composite resin on odontoblast-like cells. *J. Appl. Oral. Sci.* **2010**, *18*, 461–466. [CrossRef]
24. Kaga, M.; Seale, N.S.; Oikawa, K. Evaluation of cytotoxicity of VLC Dycal in tissue culture for clinical use. *Shoni Shikagaku Zasshi* **1989**, *27*, 313–316.
25. Chen, L.; Suh, B.I. Cytotoxicity and biocompatibility of resin-free and resin-modified direct pulp capping materials: A state-of-the-art review. *Dent. Mater. J.* **2017**, *36*, 1–7. [CrossRef] [PubMed]
26. Buonavoglia, A.; Lauritano, D.; Perrone, D.; Ardito, F.; Troiano, G.; Dioguardi, M.; Candotto, V.; Silvestre, F.J.; Lo Muzio, L. Evaluation of chemical-physical properties and cytocompatibility of TheraCal LC. *J. Biol. Regul. Homeost. Agents* **2017**, *31* (Suppl. 1), 1–9. [PubMed]
27. Kim, Y.; Lee, D.; Kim, H.M.; Kye, M.; Kim, S.Y. Biological Characteristics and Odontogenic Differentiation Effects of Calcium Silicate-Based Pulp Capping Materials. *Materials* **2021**, *14*, 4661. [CrossRef] [PubMed]
28. Calcimol LC. Available online: https://www.voco.dental/en/products/direct-restoration/liner/calcimol-lc (accessed on 21 March 2022).
29. TheraCal LC Instructions for Use. Available online: http://www.bisco.com/assets/1/22/TheraCal_LC_English4.pdf (accessed on 21 March 2022).
30. Arandi, N.Z.; Rabi, T. TheraCal LC: From Biochemical and Bioactive Properties to Clinical Applications. *Int. J. Dent.* **2018**, *2018*, 3484653. [CrossRef] [PubMed]
31. Beegum, M.S.F.; George, S.; Anandaraj, S.; Sumi Issac, J.; Khan, S.N.; Ali Habibullah, M. Comparative evaluation of diffused calcium and hydroxyl ion release from three different Indirect pulp capping agents in permanent teeth—An in vitro study. *Saudi Dent. J.* **2021**, *33*, 1149–1153. [CrossRef] [PubMed]
32. Gandolfi, M.G. A New Method for Evaluating the Diffusion of Ca(2+) and OH(−) Ions through Coronal Dentin into the Pulp. *Iran Endod. J.* **2012**, *7*, 189–197. [PubMed]
33. Luczaj-Cepowicz, E.; Marczuk-Kolada, G.; Pawinska, M.; Obidzinska, M.; Holownia, A. Evaluation of cytotoxicity and pH changes generated by various dental pulp capping materials—an in vitro study. *Folia Histochem. Cytobiol.* **2017**, *55*, 86–93. [CrossRef]
34. Didilescu, A.C.; Cristache, C.M.; Andrei, M.; Voicu, G.; Perlea, P. The effect of dental pulp-capping materials on hard-tissue barrier formation: A systematic review and meta-analysis. *J. Am. Dent. Assoc.* **2018**, *149*, 903–917. [CrossRef] [PubMed]
35. Matsuura, T.K.S.; Kawata-Matsuura, V.; Yamada, S. Long-term clinical and radiographic evaluation of the effectiveness of direct pulp-capping materials. *J. Oral. Sci.* **2019**, *61*, 1–12. [CrossRef]
36. Smaïl-Faugeron, V.; Glenny, A.M.; Courson, F.; Durieux, P.; Muller-Bolla, M.; Fron Chabouis, H. Pulp treatment for extensive decay in primary teeth. *Cochrane Database Syst. Rev.* **2018**, *5*, CD003220. [CrossRef]
37. Pereira, A.C.; Oliveira, M.D.L.; Cerqueira-Neto, A.C.C.; Gomes, B.P.; Ferraz, C.C.R.; Almeida, J.F.A.D.; Marciano, M.A.; De-Jesus-Soares, A. Treatment outcomes of pulp revascularization in traumatized immature teeth using calcium hydroxide and 2% chlorhexidine gel as intracanal medication. *J. Appl. Oral Sci.* **2020**, *28*, e20200217. [CrossRef]
38. Alqahtani, A.R.; Yaman, P.; McDonald, N.; Dennison, J. Efficacy of calcium hydroxide and resin-modified calcium silicate as pulp-capping materials: A retrospective study. *Gen. Dent.* **2020**, *68*, 50–54. [PubMed]
39. Zaparde, N.; Gunda, S.; Patil, A. Theracal ... future of pulp capping. *Int. J. Dev. Res.* **2017**, *10*, 16338–16342.
40. Voicu, G.; Didilescu, A.C.; Stoian, A.B.; Dumitriu, C.; Greabu, M.; Andrei, M. Mineralogical and Microstructural Characteristics of Two Dental Pulp Capping Materials. *Materials* **2019**, *12*, 1772. [CrossRef] [PubMed]
41. Jha, S.; Goel, N.; Dash, B.P.; Sarangal, H.; Garg, I.; Namdev, R. An Update on Newer Pulpotomy Agents in Primary Teeth: A Literature Review. *J. Pharm. Bioallied Sci.* **2021**, *13* (Suppl. 1), S57–S61.

Journal of
Functional Biomaterials

MDPI

Article

PLGA Nanoparticles Uptake in Stem Cells from Human Exfoliated Deciduous Teeth and Oral Keratinocyte Stem Cells

Maria Tizu [1], Ion Mărunțelu [1], Bogdan Mihai Cristea [2], Claudiu Nistor [3], Nikolay Ishkitiev [4], Zornitsa Mihaylova [5], Rozaliya Tsikandelova [6], Marina Miteva [4], Ana Caruntu [7,8], Cristina Sabliov [9,*], Bogdan Calenic [1,*] and Ileana Constantinescu [1]

1 Centre for Immunogenetics and Virology, Fundeni Clinical Institute, Carol Davila University of Medicine and Pharmacy, 258 Fundeni Road, 022328 Bucharest, Romania; maria.tizu@drd.umfcd.ro (M.T.); ion.maruntelu@drd.umfcd.ro (I.M.); ileana.constantinescu@imunogenetica.ro (I.C.)
2 Department of Anatomy, Carol Davila University of Medicine and Pharmacy, 8 Blvd Eroii Sanitari, 050474 Bucharest, Romania; bogdan.cristea@umfcd.ro
3 Central Military Hospital, Carol Davila University of Medicine and Pharmacy, 134 Stefan Furtuna Street, 010899 Bucharest, Romania; claudiu.nistor@umfcd.ro
4 Department of Medical Chemistry and Biochemistry, Medical Univerity of Sofia, Blvd Akademik Ivan Evstratiev Geshov 15, 1431 Sofia, Bulgaria; ishkitiev@gmail.com (N.I.); m.miteva@medfac.mu-sofia.bg (M.M.)
5 Department of Oral and Maxillofacial Surgery, Medical Univerity of Sofia, Blvd Akademik Ivan Evstratiev Geshov 15, 1431 Sofia, Bulgaria; dr.z.mihaylova@gmail.com
6 Biosciences Institute, Newcastle University, Catherine Cookson Building, Newcastle upon Tyne NE2 4HH, UK; mstsikandelova@gmail.com
7 Department of Oral and Maxillofacial Surgery, "Carol Davila" Central Military Emergency Hospital, 010825 Bucharest, Romania; ana.caruntu@gmail.com
8 Department of Oral and Maxillofacial Surgery, Faculty of Dental Medicine, "Titu Maiorescu" University, 031593 Bucharest, Romania
9 Biological and Agricultural Engineering Department, Louisiana State University and LSU Agricultural Center, 141 E. B. Doran Bldg, Baton Rouge, LA 70803, USA
* Correspondence: csabliov@agcenter.lsu.edu (C.S.); bcalenic@yahoo.co.uk (B.C.); Tel.: +1-225-578-1055 (C.S.); +40-755-044-047 (B.C.)

Abstract: Polymeric nanoparticles have been introduced as a delivery vehicle for active compounds in a broad range of medical applications due to their biocompatibility, stability, controlled release of active compounds, and reduced toxicity. The oral route is the most used approach for delivery of biologics to the body. The homeostasis and function of oral cavity tissues are dependent on the activity of stem cells. The present work focuses, for the first time, on the interaction between two types of polymeric nanoparticles, poly (lactic-co-glycolic acid) or PLGA and PLGA/chitosan, and two stem cell populations, oral keratinocyte stem cells (OKSCs) and stem cells from human exfoliated deciduous teeth (SHEDs). The main results show that statistical significance was observed in OKSCs uptake when compared with normal keratinocytes and transit amplifying cells after 24 h of incubation with 5 and 10 μg/mL PLGA/chitosan. The CD117^{+} SHED subpopulation incorporated more PLGA/chitosan nanoparticles than nonseparated SHED. The uptake for PLGA/chitosan particles was better than for PLGA particles with longer incubation times, yielding better results in both cell types. The present results demonstrate that nanoparticle uptake depends on stem cell type, incubation time, particle concentration, and surface properties.

Keywords: PLGA nanoparticles; oral keratinocyte stem cells; human exfoliated deciduous teeth

1. Introduction

PLGA or poly (lactic-co-glycolic acid), a copolymer approved by the U.S. Food and Drug Administration, is currently used in a broad range of applications in the medical field as a delivery system for biologics due to its biologic properties such as biocompatibility and biodegradability. Various therapeutic agents have been encapsulated within PLGA nanoparticles (PLGA-NP), as the polymeric matrix preserves the drugs and prevents degradation [1,2]. Depending on the tissue or system, the internalization of PLGA nanoparticles can be optimized by adjusting their size, charge, concentration, and incubation time.

In this context, oral delivery is a widely used approach for the delivery of biologics to the body, with the oral cavity being the first component of the gastrointestinal tract [3]. However, the impact of PLGA nanoparticles on oral cells, and especially oral stem cell populations, has received little attention. Therefore, their potential to act as a viable and safe delivery mechanism for oral drug administration requires further research. PLGA-NP efficiency as a carrier also depends on the oral microenvironment, which may impact the rate of nanoparticle uptake and disintegration. It is thus important to measure a baseline for which these nanoparticles are internalized by oral stem cell populations under in vitro, tightly controlled, conditions.

It is well-established that maintaining oral mucosal architecture and homeostasis is heavily dependent on the proper function of oral keratinocyte stem cells (OKSCs) [4]. Due to their importance and potential roles in regenerative medicine, several attempts have been made to isolate and characterize OKSCs. However, these cells are difficult to isolate, mainly due to the lack of appropriate surface markers. Previously, our team successfully isolated a pure subpopulation of OKSCs using two surface markers on α6β4 integrin and CD71 and a magnetic method for cell sorting. In our previous studies, we also showed that epithelial cells with the phenotype α6β4posCD71neg have important stem-cell attributes and can be successfully expanded in vitro [5].

Mesenchymal stem cells (MSCs) are already routinely separated from dental tissues. These cells are capable of, in appropriate conditions, differentiating into various cell types: adipocytes, chondrocytes, osteoblast-like cells, etc. Dental MSCs play important roles in tissue homeostasis, as they are able to repair damaged hard and soft oral structures, such as dentine and periodontal ligament [6]. To date, several types of dental-tissue-derived MSCs have been investigated: dental pulp stem cells (DPSCs), stem cells from human exfoliated deciduous teeth (SHEDs), periodontal ligament stem cells (PDLSCs), progenitor cells from dental follicles (SCDFs), and stem cells from apical papilla (SCAPs) [7,8]. Dental-derived stem cells may be the best cell source for transplantation. However, various in vitro and preclinical studies should be performed. It is well-known that the MSCs of stem cell surface markers expression include CD105, CD146, CD90, etc. In our previous studies, we successfully performed magnetic separation of DPSCs and SHEDs using a CD117 stem cell marker. Our results demonstrate that CD117+ cells possess strong potential for pancreatic differentiation in appropriate conditions, and may therefore have applications in future pancreatic disorders therapy [9].

To the best of our knowledge, nanoparticle interaction with the oral environment, particularly with oral mucosa, has been insufficiently explored. Our hypothesis in this study was that nanoparticle cellular uptake depends on several variables, such as stem cell type, length of incubation, concentration of nanoparticles, and particle surface properties. In this context, the specific aim of the present work was to assess the interactions between OKSCs, SHEDs, and PLGA nanoparticles with different attributes and to determine their uptake parameters.

2. Materials and Methods

2.1. Materials

PLGA (or poly (lactic-co-glycolic) acid at 40–75 g/mol, used 50:50), chitosan (Chi) (100–300 g/mol), polyvinyl alcohol (PVA) (31–50 g/mol) 87–89% hydrolyzed, fluorescein isothiocyanate (FITC), N-Boc-ethylenediamine, N,N-diisopropylethylamine (DIPEA),trifluoroacetic acid (TFA), 1-[Bis(dimethylamino)methylene]-1H-1,2,3-triazolo [4,5-b]pyridinium 3-oxid hexafluorophosphate) (HATU), 4′,6′-diamidino-2-phenylindole (DAPI), boric acid, and potassium iodine were acquired from Sigma-Aldrich (St. Louis, MO, USA). HPLC-grade ethyl acetate, acetone, DCM, ethanol, and acetonitrile were purchased from Mallinckrodt Baker (Pittsburgh, NJ, USA).

2.2. PLGA-FITC Conjugation

The fluorescent nanoparticles were synthesized by covalently linking FITC to PLGA. Briefly, PLGA (2 g) was dissolved in DCM (35 mL) at room temperature, and 48 mg N-Boc-ethylenediamine, 100 mg HATU, and 0.15 mL DIPEA were added under gentle stirring. After 10 h, the reaction was stopped by adding water (100 mL). PLGA-NH-Boc precipitate was obtained.

2.3. Nanoparticle Synthesis

The polymeric nanoparticles of PLGA, PLGA-FITC, and PLGA/Chi were obtained using an emulsion solvent evaporation method [10]. In short, PLGA and PLGA-FITC (400 mg) were dissolved in ethyl acetate. The solution was then added to 2% (*w/v*) 120 mL PVA aqueous solution. Ethyl acetate was further used for saturation of the aqueous phase.

The emulsion was mixed for 60 s and homogenized in a microfluidizer (M110P apparatus, Microfluidics, Westwood, MA, USA) at 4 °C and 30,000 psi for a total of 3 times. Next, the solvent was evaporated using a rotary evaporator (Buchi Inc., New Castle, DE, USA) for 1 h. In order to remove free PVA, the NPs suspension was then dialyzed with 100 kDa cut-off membrane (Spectrum Laboratories, Rancho Dominguez, CA, USA) for 48 h. Every 8 h, distilled water was replaced. The PLGA/Chi polymeric nanoparticle synthesis followed the same protocol previously described, but the sample after dialysis (10 mL) was mixed with 10 mL of 2 mg/mL Chi solution at pH 5 for 10 min. In the last step, nanoparticles were combined 1:1 (*w/w*) with trehalose and lyophilized for 2 days at −80 °C using a Labconco freeze dryer (Kansas City, MO, USA). NPs were stored at −20 °C.

2.4. Nanoparticle Characterization

NP characteristics such as size (diameter mean), polydispersity index (PDI), and zeta potential were acquired using dynamic light scattering (DLS; Zetasizer Nano, Malvern Instruments, Southborough, MA, USA). Working conditions were set as follows: sample dilution 1:20 (*v/v*) in HPLC-grade water, pH = 5, and temperature 25 °C.

Nanoparticle morphology was observed using a transmission electron microscope (TEM; JEM-1400 system, JEOL, Peabody, MA, USA). NPs were stained using 2% uranyl acetate; for analysis, the nanoparticle suspension was placed on 400-mesh carbon-coated grids.

A colorimetric assay was used to quantify the PVA remaining after nanoparticle dialysis [11]. In short, NPs powder was resuspended in 2 mL of 0.5 N NaOH and heated for 15 min at 60 °C. The samples were neutralized using 0.9 mL of 1 N HCl, and distilled water was added until a final volume of 5 mL. In the next step, 1.5 mL of distilled water, 0.5 0.05 M/0.15 M I_2/KI, and 3 mL of 0.65 M boric acid completed a volume of 10 mL. After 15 min of incubation at room temperature, the samples were measured using a spectrophotometer (Fisher Scientific, Pittsburg, PA, USA) at 690 nm. The standard curve for PVA was obtained using the same protocol for a concentration of 2.5–60 μg/mL.

2.5. Cell Culture

Isolation and separation of oral keratinocyte stem cells were performed following a protocol developed by our team and described in detail in [12]. Deciduous teeth and mucosa were supplied by patients undergoing routine deciduous tooth extraction at Dental Faculty Hospital, Medical University, Sofia, Bulgaria. All patients participating in the study signed an informed consent form following the decision of the Ethical Committee of Medical University, Sofia's Council of Medical Science (No. 4770\11.12.2018). Tissues were thoroughly rinsed for blood and debris with PBS, followed by enzymatic dissociation with 4 mg/mL dispase II (Sigma Aldrich, St. Louis, MO, USA) and 3 mg/mL collagenase (Sigma Aldrich, St. Louis, MO, USA) for 2 h at room temperature. Epidermal sheets were then removed from the dermal side, treated with trypsin 0.025% for 30 min, and centrifuged. Primary keratinocytes were cultured in EpiLife culture medium (Cascade Biologics, Portland, OR, USA) together with 1.2 mM of calcium and antibiotics—fungizone and kanamycin. The pulp was accessed through the resorbed tooth root canal and extracted by sterile barbed broach. Pulps were collected in DMEM (Invitrogen, Eugene, OR, USA) supplemented with 10% FBS (Sigma-Aldrich, St. Louis, MO, USA) and 1% antibiotic–antimycotic (Sigma-Aldrich) in sterile containers up to 24 h after extraction at 4 °C. The pulp was then digested in a PBS (Lonza, Verviers, Belgium) solution of 3 mg/mL collagenase type I/4 mg/mL dispase II for 1 h at 37 °C. Pulps were seeded in 2 cm^2 petri dishes (Greiner Bio-One, Frickenhausen, Germany) containing DMEM supplemented with 1% antibiotic–antimycotic and 20% FBS. Cell cultures from human exfoliated deciduous teeth pulp (SHEDs) were incubated in humidified atmosphere of 5% CO_2 and 95% air at 37 °C for a period of 2 to 4 weeks. The medium was removed, and a fresh one was added every second day until a monolayer of confluent cells was reached. Cells were observed using a phase contrast microscope (Leica Microsystems GmbH, Germany). After reaching 60–70% confluence, the cells were transferred to larger tissue culture flasks for further culturing. Prior to magnetic separation, the cells were expanded for 3 passages.

2.6. Magnetic Separation

Primary keratinocyte cells were further subjected to magnetic separation based on two markers expressed on the cell membrane surface: integrin α6β4 and CD71. In the first step of the separation, the cells were incubated with integrin α6β4 antibody and IgG MicroBeads. The obtained cell cocktail was placed in the magnetic field of a MACS Separator (Biotec Inc., CA, USA). Cells expressing integrin α6β4 were retained in the working column and represented the positive fraction. In the second step, this fraction was reacted with CD71 MicroBeads and subjected to the same procedure as the first isolation. Following magnetic isolation, we obtained three cell subpopulations: α6β4negative postmitotic keratinocyte cells, α6β4positive CD71positive transit amplifying cells, and α6β4 positive CD71negative fraction oral keratinocyte stem cells. Cells were routinely grown at 37 °C in 5% CO_2 in flasks precoated with collagen IV at 20 µg/mL (Sigma) and passaged at 60% confluence.

SHEDs were further subjected to magnetic separation, as described above, based on CD117 expression. The $CD117^+$ cell fraction was routinely grown at 37 °C in 5% CO_2, passaged at 60% confluence, and seeded in 24-well plates (Sigma).

2.7. Immunofluorescence

OKSCs, transit amplifying cells, and postmitotic keratinocyte cells were cultured at a density of 1×10^4 cells/well into four chamber slides (NalgeNuncInt, Naperville, IL, USA). The cells were exposed to different concentrations of PLGA and PLGA/Chi NPs: 1, 5, and 10 µg/mL for various periods of time: 3, 12, and 24 h. Following NPs exposure, each well was washed thoroughly 3–5 times with PBS and fixed in 4% paraformaldehyde, and then reacted with saturated levels of DAPI (blue) used for staining cellular nuclei. For visualization purposes, all NPs were previously conjugated with FITC (green). For semiquantitative analysis, five high-power (20× magnification) fields were randomly chosen for each slide. Cells with clearly defined staining were counted, and the count was divided by the total number of cells in each field.

Nonseparated and $CD117^+$ SHEDs seeded at a 1×10^4 cells/cm^2 density into 24-well plates were exposed to PLGA and PLGA/Chi NPs containing media at the same conditions mentioned above. After the end of the exposure time, the media containing nanoparticles was removed, cells were washed with PBS, and normal growth media without nanoparticles were introduced. The immunofluorescence images of the nanoparticle treated cells were acquired using an IN Cell Analyzer 6000 (GE Healthcare, Buckinghamshire, UK). We photographed 25 fields in each well. The green fluorescence intensity of between 1.5×10^3 and 3×10^3 cells in each well was analyzed using image analysis software (IN Cell Developer Toolbox 1.7, GE Healthcare). Cells showing values greater than 0.065 for total cytoplasmic green fluorescence were counted and compared with the total number of cells analyzed.

2.8. Statistical Analysis

The results from twenty-five fields of view for each well are presented as means ± standard deviations. Statistical analysis between different time groups and different concentration groups was performed by using one-way analysis of variance (ANOVA) (SigmaPlot 13; Systat Software GmbH, Erkrath, Germany). Statistical significance was accepted at $p < 0.05$.

3. Results

PLGA and PLGA/Chi polymeric nanoparticles were spherical in shape (Figure 1) and measured 95 nm (PLGA NPs) to 132 nm (PLGA/Chi NPs) (Table 1). The size distribution was narrow for PLGA NPs (PDI = 0.12), whereas PLGA/Chi NPs showed a wider distribution (PDI = 0.38). PLGA NPs had a negative zeta potential (−38 mV), whereas PLGA/Chi NPs were positive (+76 mV) due to the presence of chitosan on the surface (Table 1). Both particles were covered with a similar amount of PVA (25%).

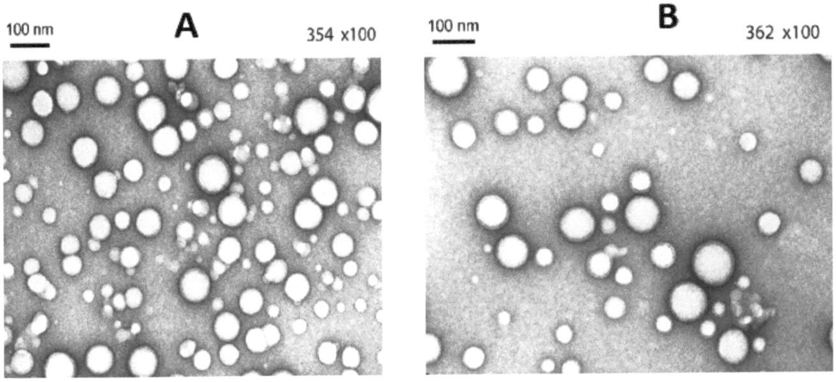

Figure 1. TEM pictures of (**A**) PLGA and (**B**) PLGA/Chi nanoparticles.

Table 1. PLGA and PLGA/Chi polymeric nanoparticles' characteristics.

	Size nm	**PI**	**Zeta Potential** mV	**PVA** %
PLGA	95 ± 2	0.11 ± 0.019	−38 ± 2	25.8 ± 0.5
PLGA/Chi	132 ± 4	0.38 ± 0.027	+76 ± 2	24.1 ± 0.8

Note: sample pH 5.5.

In the present study, three subpopulations of isolated keratinocytes and two subpopulations of SHED were exposed to 1, 5, and 10 μg/mL of PLGA and PLGA/chitosan NPs for various time intervals: 3, 12, and 24 h.

The percentage of cells containing fluorescent nanoparticles increased with exposure time and concentration of nanoparticles, but the increase was not significant, with one exception (Figure 2). A significant statistical difference was observed for OKSCs compared with NKs and TA cells, when incubated with 5 and 10 μg/mL PLGA/chitosan after one day of incubation (Figure 3A): for 5 μg/mL: 32.67 ± 2.08 vs. 9.67 ± 0.58 vs. 12.33 ± 1.53 for OKSCs, TA, and NKs, respectively; and for 10 μg/mL: 27.67 ± 1.53 vs. 9.63 ± 1.53 vs. 10.52 vs. 1.53 for OKSCs, TA, and NKs, respectively (data presented as means ± SD; $p < 0.05$. ANOVA, n = 5 independent experiments). Similarly, no statistical increases were detected in keratinocyte cells incubated with PLGA NPs at concentrations of 1, 5, and 10 μg/mL for 3, 12, and 24 h (Figure 3B).

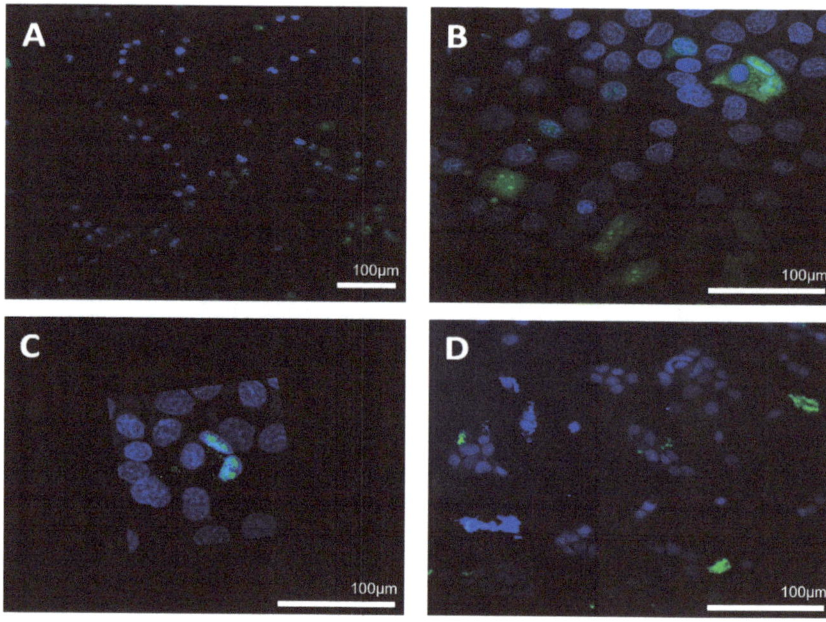

Figure 2. Immunofluorescence: oral keratinocyte stem cells uptake of PLGA/chitosan nanoparticles following 24 h exposure (blue DAPI staining was used for nuclei, while NPs are labelled with FITC and appear in green): (**A**) ×20 magnification; (**B**) ×40 magnification; (**C**) 3D model; (**D**) NPs outside cell membrane; nanoparticle concentration—5 μg/mL.

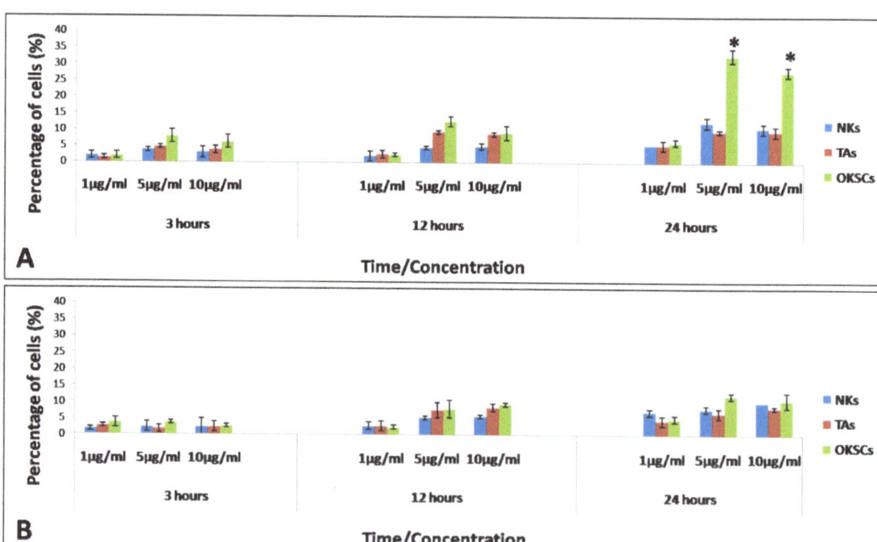

Figure 3. Uptake of polymeric nanoparticles by different oral keratinocyte cells at various NPs concentrations and different time points. (**A**) PLGA/chitosan uptake by normal keratinocytes (NKs), transit amplifying cells (TA), and oral keratinocyte stem cells (OKSCs); (**B**) PLGA uptake by normal keratinocytes (NKs), transit amplifying cells (TAs), and oral keratinocyte stem cells (OKSCs). Statistical significance was set at * $p < 0.05$ ANOVA; n = 5 independent experiments.

The uptake of nanoparticles was monitored by immunofluorescence (Figure 4). Analysis of the SHED cells uptake of PLGA (Figure 5A uptake graph) showed around 60% to 73% of nonseparated cells uptake of nanoparticles after 3 h of incubation. Over time, the percentage of fluorescent cells increased to about 75% of the cells, and, until the end of the incubation at 24 h, remained stable with one exception—the concentration of 1 μg/mL dropped back to 64%. Most stable was the uptake of 5 μg/mL PLGA NPs after 3 h of incubation at 73.37% ± 12.11%, 12 h of incubation at 74.67% ± 7.97%, and 24 of incubation at 74.70% ± 5.00%.

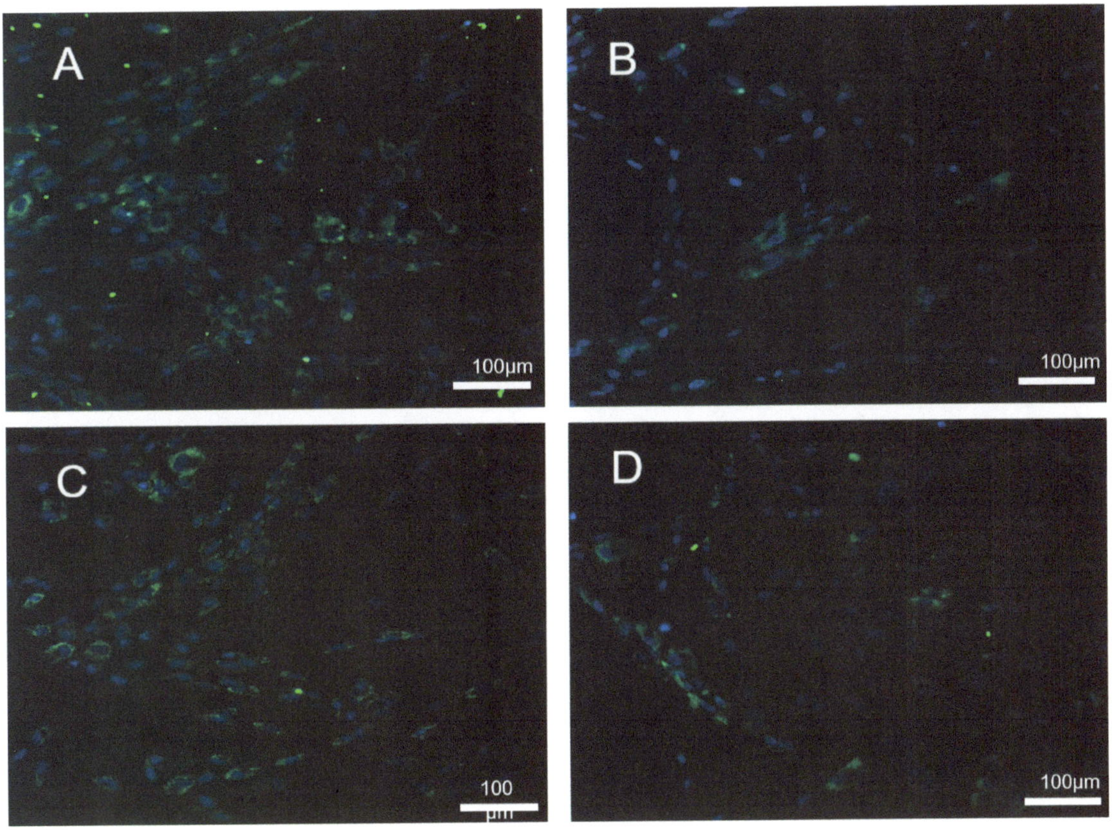

Figure 4. Immunofluorescence: SHED uptake of PLGA and PLGA/chitosan nanoparticles following 24 h exposure (blue DAPI staining was used for nuclei, while NPs are labeled with FITC and appear in green). (**A**) Normal SHED PLGA uptake; (**B**) CD117+ SHED PLGA uptake; (**C**) normal SHED PLGA/chitosan uptake; (**D**) CD117+ SHED PLGA/chitosan uptake. Magnification ×20; nanoparticle concentration 5 μg/mL.

A statistically significant difference was found in the $CD117^+$ fraction uptake compared with the same conditions for the nonseparated SHEDs. The percentage of cells uptake for PLGA NPs remained stable around 95% throughout the entire incubation for all concentrations. Nonseparated SHEDs expressed similar uptake rates for the PLGA/chitosan NPs (Figure 5B uptake graph). Between 66% and 75% of the cells showed green fluorescence. The uptake of 1 and 5 μg/mL PLGA/chitosan NPs dropped at 24 h compared with 3 h of incubation, respectively, to 62.15% ± 9.43% from 66.35% ± 6.41% and to 63.24% ± 14.6% from 75.45% ± 9.43%, as the uptake of 10 μg/mL increased from 70.40% ± 13.76% to 75.93% ± 10.17%. The $CD117^+$ fraction showed 91.76% ± 5.35% at

1 µg/mL; 90.67% ± 6.39% at 5 µg/mL and 96.10% ± 2.55% at 10 µg/mL after 3 h of incubation with PLGA/chitosan NPs. At the end of the incubation, the number of fluorescent cells increased to 93.55% ± 3.10% at 1 µg/mL and 93.75% ± 4.67% at 5 µg/mL, and dropped to 94.65% ± 3.26% at 10 µg/mL. Significantly more cells were showing NPs uptake from CD117$^+$ compared with the nonseparated SHEDs.

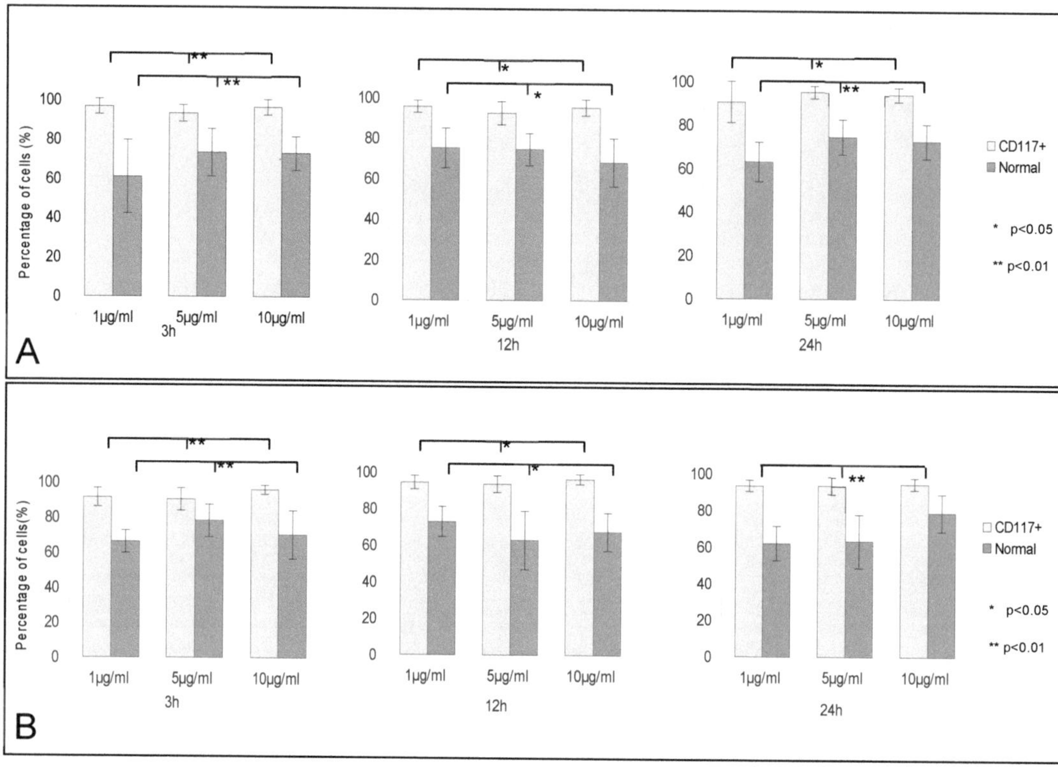

Figure 5. Uptake of polymeric nanoparticles by normal and CD117+ SHEDs at various NP concentrations and different time points. (**A**) PLGA uptake by normal and CD117+ SHEDs; (**B**) PLGA/chitosan uptake by normal and CD117+ SHEDs. Statistical significance was set at * $p < 0.05$ and ** $p < 0.01$ ANOVA; n = 5 independent experiments.

4. Discussion

One major challenge when addressing stem cell therapy is finding adequate scaffolds or careers knowing that tissue regeneration requires an optimal combination of cells and active molecule growth factors. Therefore, a wide range of organic and inorganic materials are currently applied in various in vitro, preclinical, and clinical studies. A novel method for active substances in tissue regeneration is the use of nanoparticles. Among these, polymeric NPs have recently been introduced as drug delivery vehicles, as their application has resulted in improved pharmaceutical bioavailability, enhanced efficacy, and reduced toxicity [13]. Poly (d,l-lactide-coglycolide acid) (PLGA) is a widely investigated polymer for NP synthesis [14]. It is able to improve the oral bioavailability of DNA and peptides [15]. Chitosan is another biocompatible nontoxic polymer. It is an N-deacetylated derivate of chitin known for its ability to promote sustained release of various proteins, hydrophilic and hydrophobic drugs, and DNA transfection [16]. Thus, PLGA nanoparticles are broadly accepted drug carriers due to their biocompatibility and ability to provide sustained, controlled, and targeted drug delivery.

The body recognizes particles as foreign bodies, and the reticulo-endothelial system (RES) eliminates them from the blood. Additional surface modification, i.e., PEGylation, can increase PLGA nanoparticles' half-life and prolong their circulation in the blood stream.

Passive transport and active endocytosis are well-known methods for cellular uptake of nanoparticles. Two main types of endocytosis have been described in the literature: nonspecific endocytosis and receptor-mediated endocytosis, which includes clathrinid-mediated endocytosis and caveolae-mediated endocytosis [17]. The pathway of cellular uptake depends on the surface characteristics and size of the nanoparticles. Preparation conditions are of high importance for the biological applications of nanoparticles. Surface charges of nanoparticles affect their cellular internalization. Positively charged nanoparticles are expected to have more sustained cellular uptake due to the positively charged surface of the nanoparticle and negatively charged cell membrane. PLGA nanoparticles are known to have negative charges as well as the cell surface. The particles' negative charge can be shifted to neutral, or even positive, by surface modification, including PEGylation (polyethylene glycol (PEG) or chitosan coating [18]. It is well-known that chitosan also possesses absorption-enhancing properties and is able to accelerate cell uptake.

Within the past few years, mesenchymal stem cells have been widely discussed as a potential cancer gene therapy vehicle [19]. The combination of stem cells and loaded particles may lead to even better results in anticancer therapy. In the present study, the cellular uptake of fluorescent nanoparticles by OKSCs and SHEDs was identified by semiquantitative immunofluorescence. PLGA and PLGA/chitosan nanoparticles were not loaded with active molecules and drugs. The nanoparticles in the current experiment were spherical in shape with a smooth surface. The aim was just to assess the differences in cellular uptake after stem cells' incubation with various concentrations and durations of incubation time. We included two different fractions of SHED cells in the experiment: nonseparated SHEDs and a CD117+ fraction. CD117 is a well-known stem cell marker, previously applied in in vitro studies for magnetic and fluorescent activated cell sorting. The marker is expressed in wide range stem cell cultures. Our results revealed that 5 and 10 μg/mL PLGA uptake remained stable in both cell fractions at the 3rd, 12th, and 24th hour of incubation. When these concentrations are applied, the uptake is not time dependent. For the nonseparated SHEDs incubated with 1 μg/mL PLGA, the uptake was significantly increased at the 12th hour of incubation, whereas for the CD117+ fraction in the same conditions, the uptake decreased. The same tendency was observed in cells incubated with 1 μg/mL PLGA/chitosan. In the normal nonseparated SHED fraction, the uptake significantly decreased at the 12th hour of incubation with 5 and 10 μg/mL PLGA/chitosan. The optimal PLGA and PLGA/chitosan nanoparticles' uptake for both SHED cell fractions occurred at a concentration of 1 μg/mL; the optimal PLGA and PLGA/chitosan incubation time for nonseparated cells was 3 h, and for the CD117+ cells, it was 12 h. Therefore, the cell uptake mechanism depends on the cell fraction, incubation time, and surface properties of the nanoparticles.

The present study established, for the first time, a baseline for PLGA-NP uptake by two subpopulations of oral stem cells, oral keratinocyte stem cells, and dental stem cells. We report important uptake parameters such as particle size, particle chemical composition, time, and concentration tailored for each studied cell population. However, more studies are needed to certify these parameters, not only in an in vitro setting, but also in vivo.

Author Contributions: Supervision, I.C., B.C. and C.S.; writing—original draft preparation, M.T., I.M. and C.N.; writing—review and editing, N.I., C.S., B.C. and I.C.; methodology writing review and editing B.M.C.; methodology, Z.M., R.T., M.M. and A.C.; conceptualization, C.S., B.C., and I.C.; formal analysis, N.I., A.C. and B.C. All authors have read and agreed to the published version of the manuscript.

Funding: This research was partially funded by a Fulbright Senior Grant titled: Polymeric nanoparticle synthesis targeted for oral epithelial stem cells.

Institutional Review Board Statement: The study conducted in accordance with the Declaration of Helsinki. All patients participating in the study signed an informed consent form following the decision of the Ethical Committee of Medical University, Sofia's Council of Medical Science (No. 4770\11.12.2018).

Informed Consent Statement: Informed consent was obtained from all subjects involved in the study.

Data Availability Statement: Not applicable.

Conflicts of Interest: The authors declare no conflict of interest. The funders had no role in the design of the study; in the collection, analyses, or interpretation of data; in the writing of the manuscript; or in the decision to publish the results.

References

1. Martins, C.; Sousa, F.; Araújo, F.; Sarmento, B. Functionalizing Plga and PLGA Derivatives for DrugDelivery and Tissue Regeneration Applications. *Adv. Healthc. Mater.* **2017**, *7*, 1701035. [CrossRef] [PubMed]
2. Virlan, M.J.R.; Miricescu, D.; Totan, A.; Greabu, M.; Tanase, C.; Sabliov, C.M.; Constantin, C.; Calenic, B. Current uses of poly (lactic-co-glycolic acid) in the dental field: A comprehensive review. *J. Chem.* **2015**, *2015*, 525832. [CrossRef]
3. Cao, S.J.; Xu, S.; Wang, H.M.; Ling, Y.; Dong, J.; Xia, R.D.; Sun, X.H. Nanoparticles: Oral delivery for protein and peptide drugs. *Aaps Pharmscitech* **2019**, *20*, 190. [CrossRef] [PubMed]
4. Calenic, B.; Greabu, M.; Caruntu, C.; Tanase, C.; Battino, M. Oral keratinocyte stem/progenitor cells: Specific markers, molecular signaling pathways and potential uses. *Periodontology 2000* **2015**, *69*, 68–82. [CrossRef] [PubMed]
5. Calenic, B.; Ishkitiev, N.; Yaegaki, K.; Imai, T.; Costache, M.; Tovaru, M.; Parlatescu, I. Characterization of oral keratinocyte stem cells and prospects of its differentiation to oral epithelial equivalents. *Rom. J. Morphol. Embryol.* **2010**, *51*, 641–645. [PubMed]
6. Sharpe, P.T. Dental mesenchymal stem cells. *Development* **2016**, *143*, 2273–2280. [CrossRef] [PubMed]
7. Akiyama, K.; Chen, C.; Gronthos, S.; Shi, S. Lineage differentiation of mesenchymal stem cells from dental pulp, apical papilla, and periodontal ligament. In *Odontogenesis*; Humana Press: Totowa, NJ, USA, 2012; pp. 111–121.
8. Estrela, C.; Alencar, A.H.G.D.; Kitten, G.T.; Vencio, E.F.; Gava, E. Mesenchymal stem cells in the dental tissues: Perspectives for tissue regeneration. *Braz. Dent. J.* **2011**, *22*, 91–98. [CrossRef] [PubMed]
9. Ishkitiev, N.; Yaegaki, K.; Kozhuharova, A.; Tanaka, T.; Okada, M.; Mitev, V.; Fukuda, M.; Imai, T. Pancreatic differentiation of human dental pulp $CD117^+$ stem cells. *Regen. Med.* **2013**, *8*, 597–612. [CrossRef] [PubMed]
10. Zigoneanu, I.G.; Astete, C.E.; Sabliov, C.M. Nanoparticles with entrapped α-tocopherol: Synthesis, characterization, and controlled release. *Nanotechnology* **2008**, *19*, 105606. [CrossRef] [PubMed]
11. Murugeshu, A.; Astete, C.; Leonardi, C.; Morgan, T.; Sabliov, C.M. Chitosan/PLGA particles for controlled release of α-tocopherol in the GI tract via oral administration. *Nanomedicine* **2011**, *6*, 1513–1528. [CrossRef] [PubMed]
12. Calenic, B.; Ishkitiev, N.; Yaegaki, K.; Imai, T.; Kumazawa, Y.; Nasu, M.; Hirata, T. Magnetic separation and characterization of keratinocyte stem cells from human gingiva. *J. Periodontal Res.* **2010**, *45*, 703–708. [CrossRef] [PubMed]
13. Begines, B.; Ortiz, T.; Pérez-Aranda, M.; Martínez, G.; Merinero, M.; Argüelles-Arias, F.; Alcudia, A. Polymeric nanoparticles for drug delivery: Recent developments and future prospects. *Nanomaterials* **2020**, *10*, 1403. [CrossRef] [PubMed]
14. Ghitman, J.; Biru, E.I.; Stan, R.; Iovu, H. Review of hybrid PLGA nanoparticles: Future of smart drug delivery and theranostics medicine. *Mater. Des.* **2020**, *193*, 108805. [CrossRef]
15. Gomes dos Reis, L.; Lee, W.H.; Svolos, M.; Moir, L.M.; Jaber, R.; Windhab, N.; Young, P.M.; Traini, D. Nanotoxicologic effects of PLGA nanoparticles formulated with a cell-penetrating peptide: Searching for a safe pDNA delivery system for the lungs. *Pharmaceutics* **2019**, *11*, 12. [CrossRef] [PubMed]
16. Jhaveri, J.; Raichura, Z.; Khan, T.; Momin, M.; Omri, A. Chitosan nanoparticles-insight into properties, functionalization and applications in drug delivery and theranostics. *Molecules* **2021**, *26*, 272. [CrossRef] [PubMed]
17. Manzanares, D.; Ceña, V. Endocytosis: The nanoparticle and submicron nanocompounds gateway into the cell. *Pharmaceutics* **2020**, *12*, 371. [CrossRef] [PubMed]
18. Danhier, F.; Ansorena, E.; Silva, J.M.; Coco, R.; Le Breton, A.; Préat, V. PLGA-based nanoparticles: An overview of biomedical applications. *J. Control. Release* **2012**, *161*, 505–522. [CrossRef] [PubMed]
19. Lin, W.; Huang, L.; Li, Y.; Fang, B.; Li, G.; Chen, L.; Xu, L. Mesenchymal stem cells and cancer: Clinical challenges and opportunities. *BioMed Res. Int.* **2019**, *2019*, 2820853. [CrossRef] [PubMed]

MDPI AG
Grosspeteranlage 5
4052 Basel
Switzerland
Tel.: +41 61 683 77 34

Journal of Functional Biomaterials Editorial Office
E-mail: jfb@mdpi.com
www.mdpi.com/journal/jfb

www.ingramcontent.com/pod-product-compliance
Lightning Source LLC
Chambersburg PA
CBHW042009260926
51850CB00020B/2939
* 9 7 8 3 7 2 5 8 4 2 0 5 6 *